Electronics Pocket Book

Books of Allied Interest

BEGINNER'S GUIDE TO COLOUR TELEVISION
by Gordon J. King

BEGINNER'S GUIDE TO ELECTRONICS
by T. L. Squires and C. M. Deason

BEGINNER'S GUIDE TO TRANSISTORS
by J. A. Reddihough

INTEGRATED CIRCUIT POCKET BOOK
by R. G. Hibberd

RADIO SERVICING POCKET BOOK
by Vivian Capel

QUESTIONS AND ANSWERS ON ELECTRONICS
by Clement Brown

TRANSISTOR POCKET BOOK
by R. G. Hibberd

Electronics Pocket Book

Edited by

P. J. McGOLDRICK, C.Eng., M.I.E.E., M.S.M.P.T.E.

Newnes Technical Books

THE BUTTERWORTH GROUP

ENGLAND	Butterworth & Co (Publishers) Ltd London: 88 Kingsway, WC2B 6AB
AUSTRALIA	Butterworths Pty Ltd Sydney: 586 Pacific Highway, NSW 2067 Also Melbourne, Brisbane, Adelaide and Perth
CANADA	Butterworth & Co (Canada) Ltd Scarborough: 2265 Midland Avenue, Ontario M1P 4S1
NEW ZEALAND	Butterworths of New Zealand Ltd Wellington: 26–28 Waring Taylor Street, 1
SOUTH AFRICA	Butterworth & Co (South Africa) (Pty) Ltd Durban: 152–154 Gale Street

First published in 1963 by George Newnes Ltd
Second edition 1967
Third edition 1976, published by Newnes Technical Books, a Butterworth imprint

ISBN 0 408 00209 3

Filmset and printed in England by
J. W. Arrowsmith Ltd., Bristol BS3 2NT

PREFACE

In the thirteen years since the first edition of this pocket book the developments in the world of electronics have been almost unbelievable. The semiconductor device in all its variants is an established part of life and the convenience of size, low power consumption and simplicity have corrupted our minds into thinking that electronics means small. In many instances this may be so but there is still a need for some thermionic devices, albeit specialised ones, in the industrial control and r.f. fields—where the semiconductor is still operationally limited or uneconomic.

For that reason, although the emphasis in this third edition is on semiconductors, information on the thermionic device has not been discarded completely. The semiconductor establishes its place with a mass of new material which, I hope, will describe the operation and theory of many devices in as simple a manner as possible.

The philosophy of this book is simple: to bring together as much as possible of the present-day knowledge of basic circuits and techniques in electronics, in a practical way, without getting involved in the mathematics of the processes concerned. While no claims are made for the coverage of any particular examination syllabus, the electronic technician and the hobbyist should both find the book of value.

The first two editions were edited by J. P. Hawker and J. A. Reddihough; their commitments restrain them from further involvement but I am grateful to them both for establishing the pocket book as an invaluable reference text, making it even more pleasurable for me to take the responsibility of producing this new edition.

P. J. McGoldrick

CONTENTS

1
THE NATURE OF THE ELECTRON

Electronics, as the name implies, is a subject which deals with apparatus and phenomena in which the behaviour and properties of electrons play major roles. It is therefore of fundamental importance for the understanding of electronics to know what electrons are, where they are to be found and how they behave under the conditions likely to be met with in electronic equipment.

The Structure of the Atom

Experimental evidence from work on the photo-electric effect and on gas discharge phenomena suggested from the first that neutral atoms and collections of atoms contained equal quantities of positive and negative charges. Modern views on the distribution of these charges within the atomic structure are based on the Rutherford–Bohr model of the atom. The atom is built up of a core, or nucleus, of positive and neutral particles surrounded by a cloud of negatively charged particles. The positive, neutral and negative particles are identical in the atoms of all materials, the only difference between the atoms of different elements being the number of these particles which combine to form the atom.

The existence of the negative particles was discovered in 1897 by Sir J. J. Thomson, who called them electrons. They are fundamental constituents of all matter and are found throughout the universe. The electron is the smallest of the so-called fundamental particles, having a mass of $0{\cdot}9107 \times 10^{-27}$ g, and carrying a negative charge of 16×10^{-20} coulomb or $4{\cdot}8 \times 10^{-10}$ e.s.u. Although they are a part of all atoms electrons can exist in a free state. The electrons are held to the nucleus by forces which can be thought of as being mainly the forces of attraction between positive and negative charges. The retaining forces for some of the electrons are weaker than for others, and when atoms collide with other atoms or with fast-moving particles some of these loosely bound electrons may be knocked off and drift away on their own. In the course of time a free electron will meet an atom which has some electrons missing from its structure and the free electron will be captured. These processes of free electron production and capture are going on continuously, so that a balance is struck between them, which leaves a continual population of free electrons.

This method of producing free electrons seems rather haphazard, and ways in which free electrons can be produced to order will be considered later.

BEHAVIOUR OF ELECTRONS IN ELECTRIC AND MAGNETIC FIELDS

The earth's gravitational field does have an effect on the motion of an electron, but the mass of the electron is so small that gravitational effects cannot be detected with present-day techniques and can therefore be neglected in electronic applications. The electronic charge, however, produces considerable interactions with electric and magnetic fields.

The Electric Field and Potential

It is a well-known experimental fact that if two electrically charged bodies are brought near to each other they will be acted on by easily measurable mechanical forces which are due to the existence of the charges. If the charges are of the same sign, that is both positive or both negative, the charged bodies will repel each other. If the charges are of different sign—one positive and one negative—then the forces acting on the bodies will try to pull them together. If an electron is placed in the region between the two charged bodies, it will be attracted to the positively charged one and repelled by the other one: the electron will therefore move towards the positive charge.

It will now be useful to bring in two new ideas: electric field strength and potential. An electric field is said to exist at any point if a charge placed at that point has a mechanical force exerted on it because of that charge. If the sign of the charge is altered the direction of the force is reversed. If there is no charge there will be no force.

At each point in a region which contains charged bodies a line is drawn parallel to the direction of the mechanical force acting on a small test charge (a small charge is used to avoid distorting the field) placed at that point. Curves can then be drawn such that the direction of the force at each point on a curve is a tangent to the curve. A charge placed in the field and allowed to move under the action of the field forces will follow one of these curves, or field lines. In practice this method of mapping an electric field is seldom used because theoretical considerations reveal more sophisticated and accurate ways of finding field strengths and directions. Nevertheless this idea of introducing a probe into a field in order to discover the field strength and direction can be quite valuable on occasions. The electrolytic tank is an example of this technique.

Now that the field has been mapped, the charged bodies which gave rise to the field can be left out of the picture. Effects in the field can be determined from a knowledge of the field lines only.

According to Coulomb's Law, the force between two charges is proportional to the product of the charges. Since the field is due to charges, the force which a field exerts on a charged body such as an electron will be proportional to the charge on that body. The quantity by which the charge must be multiplied to obtain the force on the charge is the electric field strength at the point in the field at which the charge is located. Once the position and size of all the charges which are giving rise to the field are known, their resultant field can be obtained for all points in the field and, since the force between charges depends on the distance between the charges, the field strength throughout the region will not in general be a constant quantity:

for example, the field close to a charged body would be expected to be higher than at a position a little further away.

When the point of application of a force moves through a distance in the direction of the force, the force is said to do work. The work done is given by the product of the force and the distance moved in the corresponding direction. If the force varies from point to point, it is necessary, to obtain the correct value, to determine the work done over successive short distances, short enough that the change in the force is negligible, and then add all these small contributions together to obtain the total work done.

The work done by an electric force when a charge moves under the force in an electric field can be obtained by multiplying the field strengths along the path by the path length. The work which has to be done to move a unit charge from a point A to another point B in the field is called the potential difference (or voltage) between A and B.

This gives another form of mapping of the field. Any movement of a charge along a path at right-angles to the field lines at each point will not, by definition, require any work on the part of the field. Another set of lines can now be drawn which are everywhere at right-angles to the field lines. This second set of lines is known as the set of equi-potential lines, for there is no change in potential on moving along any one of these lines.

These equi-potential lines are much the same as contour lines on a terrestrial map. Contour lines join points which have the same height: equi-potential lines join points which have the same potential. The potential difference between two points is given by the product of charge, distance and field strength. Where the potential lines are close together the field strength must be greater than in regions where the lines are further apart. Just as the closeness of contour lines gives the steepness of a slope, so the closeness of equi-potential lines gives the strength of the field. Field lines give the directions which would be taken by a charge if it were released at the top of a potential slope and allowed to roll to the bottom.

Positive charges are customarily regarded as being regions of high potential, so that continuing the analogy with a map a solitary positive charge would be like a lone hill set in a plain.

Motion of an Electron in an Electric Field

These concepts of field and potential can now be used to determine what happens to an electron which is placed in an electric field. Take two large flat metal plates placed in parallel planes (Fig. 1). A voltage is applied between the plates. A little

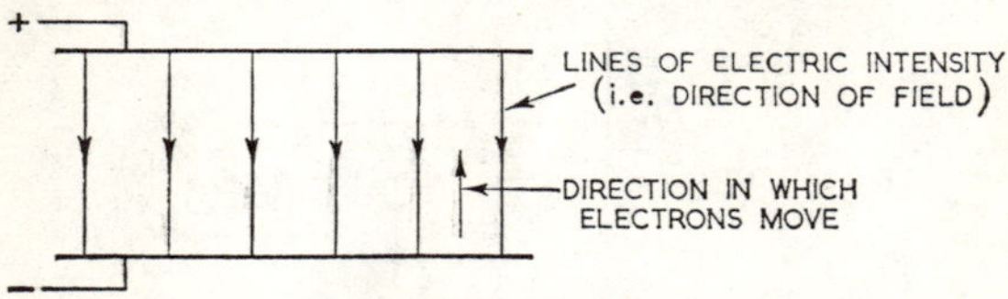

Fig. 1.—*Electric field between two parallel charged plates.*

thought will show that the field lines will be parallel straight lines running from plate to plate and at right-angles to the plates, while the equi-potential lines will be straight lines parallel to the plates. An electron set between the plates will move along the field line through the point where it was set until it reaches the positive plate. (Note that because it has a negative charge it moves "up" the potential hill.) As it moves it will increase in speed and energy.

If the electron introduced into the field has an initial speed parallel to the field it will continue to move in a straight line in the same direction. Its path will not be deflected. See Fig. 2.

Thus it can be seen that an electric field whose direction is parallel to that of a moving electron can be used to speed up an electron, or slow it down if the field is in the opposite direction. This is greatly used in electronic equipment. In the cathode-ray tube electrons are produced at a source called a cathode and are allowed to fall on a screen which fluoresces under the impact of the electrons. By placing an electrode which carries a high voltage between the cathode and the screen, an electric field is produced which draws the electrons towards this "accelerating" electrode. A hole in this electrode allows the speeded-up electrons to pass through it and fall on the screen. The greater the voltage on this electrode the greater the speed with which the electrons reach the screen and the brighter the image formed on the screen.

In thermionic valves the electrons produced at the cathode of a valve are drawn to another electrode (the anode) by a positive voltage on the anode. As this voltage is increased the

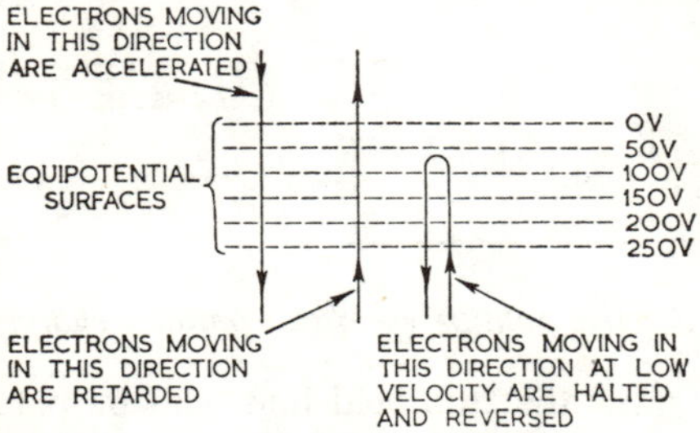

Fig. 2.—*Behaviour of an electron beam entering an electric field at right-angles to equipotential surfaces.*

number of electrons attracted to the anode increases. A grid of fine wires placed between anode and cathode will still allow all but a few electrons to pass through the meshes of the grid, but if now the grid is given a negative voltage the electrons will have to "climb" a potential hill after they leave the cathode before they can reach the anode. Some of the electrons fail to get over the hill. By varying the voltage on the grid, it is possible to permit more or fewer electrons to pass through the grid and reach the anode. The implications of the control of electron flow are of great importance in valve operation.

Field Perpendicular to the Motion of the Electron

Suppose that an electron is injected into the space between the two metal plates so that it is moving parallel to the plates and so at right-angles to the field. Its forward velocity will be unaltered since the forces on the electron are at right-angles to this direction. But as it traverses the space between the plates it will be pushed sideways by the field and move closer to the positive plate. Its motion will be similar to that of a ball rolled horizontally across a sloping table: it will tend to roll down the slope. Its path is deflected.

This effect is made use of in cathode-ray tubes (Fig. 3). The beam of electrons on its way to the screen passes between two

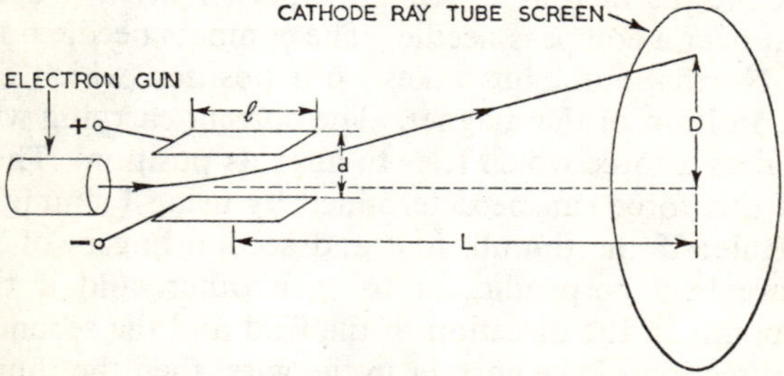

Fig. 3.—*Electrostatic deflection of an electron beam in a cathode-ray tube. Deflection* D *on the screen is related to the voltage* V *between the plates according to the approximate expression*

$$\frac{D}{V} = \frac{lL}{2dV_0},$$

where l is the length of the plates, d *the spacing between them,* L *the distance from plates to screen and* V_0 *the final accelerating potential of the electron gun.*

metal plates which give a field at right-angles to the direction of the beam. When no voltage is applied to these plates no field is present and so the electron beam is undeflected. If a voltage is applied to the plates the beam will be deflected and will now strike the screen at a different point. The greater the voltage on the plates, the greater the force deflecting the beam and the greater the deflection of the beam. By measuring just how much the beam has been deflected it is possible to determine what voltage has been applied to the plates. Many instances of the use of "deflection plates" are met with in electronic equipment.

Behaviour of Electrons in Magnetic Fields

Magnets give rise to fields whose field lines and lines of equal magnetic potential can be plotted just as in the case of electric fields. Magnetic fields are also set up by a wire which is carrying a current, and if this wire is then placed in a magnetic field due to some other source the two fields interact. This is the interaction which can be seen when a magnet is brought near a compass needle. The compass needle no longer points North–South, but takes up a position which depends on the position of the magnet. The current-carrying wire also experiences a force which tries to alter its position. The direction of this force can be determined by using Fleming's Left-hand Rule: If the thumb, first and second fingers of the left hand are held perpendicular to each other, and if the first finger points in the direction of the field and the second finger in the direction of the current in the wire, then the thumb will be pointing in the direction of the mechanical force on the wire. (It is this phenomenon which is the principle of the electric motor.)

A moving charge acts like a current in a wire, so that if an electron moves in a magnetic field its subsequent motion can be determined by considering it to be part of a current-carrying wire. Consequently an electron which is moving at right-angles to a magnetic field will be acted on by a force which is at right-angles to its path at all times (Fig. 4). Note carefully here the difference between this case and that of the electric field. In the electric field the force was at right-angles to the electron's original path: in the magnetic field it is always at right-angles to the electron's path. This means that in a magnetic field the path of an electron moving in a plane perpendicular to a magnetic field will be a circle, just as a weight swung round on a length of string moves in a circular path under the influence of the tension in the string.

This spiralling effect is used in the magnetic focusing coils of a cathode-ray tube. As the beam of electrons moves towards the screen the beam tends to spread out, partly because not all the electrons have quite the same direction, and partly because the electrons are forced apart by the mutually repulsive effect of their charges. This spreading tendency is reduced by means of a magnetic field produced either by a permanent magnet or, more conveniently, by coils through which a current is passed. The shape of the field is designed so that an electron which

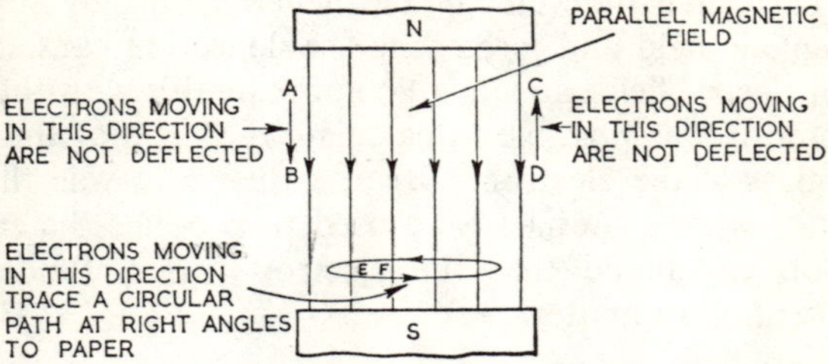

Fig. 4.—*The forces acting on electrons moving in a parallel magnetic field.*

is travelling along the axis of the tube is not affected by the field but any electrons which are drifting out of line are forced back by the field into the correct direction.

Behaviour of Electrons in Combined Electric and Magnetic Fields

A combination of electric and magnetic fields is used to produce high energy charged particles in such machines as synchrotrons. As was shown earlier, it is possible to speed up charged particles by accelerating them between two electrodes at different potentials. The maximum energy that it is possible to transfer to the particles in this way is limited by the greatest voltage difference that it is possible to obtain. If the whole system is placed in a magnetic field, the paths of particles will be circular and can be made to traverse the voltage difference time after time: each time a particle goes through the voltage gap it will be further accelerated, and so much higher energies can be obtained by these means.

If an electron passes through a region which contains an electric field and a magnetic field, and if the fields and the electron's path are all perpendicular to each other, then it can be seen that the directions of the electric and magnetic forces on the electron will be the same. Sir J. J. Thomson used this arrangement of field in his work. The strengths. of the fields were adjusted so that the deflection of the electron by one field was exactly cancelled out by the deflection due to the other field. This gave the velocity of the electron. By removing the magnetic field and measuring the deflection caused by the original electric field alone, and combining this with the previous result, it was possible to calculate the ratio of the charge to the mass of the electron. Because this ratio was the same no matter what materials were used to produce the electron, Thomson concluded that the electron was a fundamental constituent of all matter.

Ions

An atom or molecule that has lost one or more electrons is called a positive ion. Its charge is therefore equal to the negative charge of the lost electron(s) while its weight approximates to that of the original atom or molecule. Ions therefore differ from electrons, firstly because they are much heavier and secondly because they are chemically the same as the original atom or molecule—e.g. hydrogen ions or mercury ions.

Ions are produced when a charged body—either an electron moving rapidly or a previously produced ion—strikes a molecule of gas and dislodges an electron from it. With a net resultant positive charge, the newly formed ion will be attracted towards a negative potential—for example, a cathode. This process is known as *ionisation by collision.*

Energy Levels

In an atom each electron associated with the nucleus must lie in one of the orbits that are clearly defined by quantum mechanics. It requires energy to move an electron from one orbit to another outer orbit, so each level represents a degree of potential energy.

An electron will therefore take up position in the innermost orbit that has space for it (there being, for each orbit, a maximum possible number of electrons) so that the lower energy levels are occupied while the higher levels remain unoccupied.

Crystalline Solids

A number of solids exist that are obviously crystalline in their structure; *crystalline* is defined as having a regular appearance with faces in which corresponding faces have the same angle between them. In fact there are surprisingly few materials that are non-crystalline.

Equilibrium—a minimum, balanced energy state—demands that atoms have an ordered arrangement; to comply with this law atoms are neatly arranged in rows, the rows laid together to form planes and the planes stacked to form an ordered three-dimensional block. Those materials that do not appear to the naked eye to be crystalline will do so under a microscope. The exceptions are those few materials that have a poorly discernible transition between solid and liquid under heating—for example, glass and plastic.

CONDUCTION AND CURRENT-VOLTAGE RELATIONSHIPS

The free electrons which are present in such materials as metals drift around in the metal: it is as if the metal were impregnated with a negatively charged gas. If a voltage is applied across the ends of a piece of metal, the electric field causes the electrons to move through the metal. An electric current is said to flow, and the material is called a conductor.

It can be shown experimentally that when a voltage is applied to a typical conductor the current which flows is proportional to the voltage producing it. That is, $V/I =$ a constant, R, and this constant R is called the resistance of the conductor. As the electrons are drawn through the conductor they collide with the atoms present, give up some of the energy and speed that they have gained from the field and slow down, only to be accelerated again by the field until another collision occurs. This loss of energy by the electron gas to the material is equivalent to a frictional retarding force which prevents unlimited speeding up of the electrons, and so maintains a constant relationship between the current and voltage. The energy which is passed on to the atoms shows itself in the heating up of the metal—known as Joule heating.

This original simple theory of electron conduction in metals had some initial success and is still useful in many respects. It helps to explain why the resistance of metals rises as the temperature rises. The hotter the metal the greater the agitation of its atoms and so the greater the number of atom–electron collisions which occur and the more difficulty the electrons have in moving through the metal. But when this theory was applied to other effects it was unable to explain them satisfactorily and has to a great extent been superseded by more sophisticated ideas.

Materials in which the current-proportional-to-voltage relationship (or Ohm's Law) is obeyed are termed ohmic conductors. Most solids are ohmic. In the basic equation $V = IR$, R is in ohms, I amperes and V volts.

Some materials show deviations from Ohm's Law because of the Joule heating effect: passage of current causes the temperature of the material to rise and so its resistance alters. Higher currents cause more heating and so a greater change in resistance.

Some other materials, such as the ceramic thyrite, obey a non-ohmic relationship of the form $I = kV^n$, where k is a constant depending on composition and dimensions, and n lies between 3 and 5.

Conduction

Conduction in solids is an electronic process. Conduction in liquids is generally ohmic: the current carriers are rarely electrons, but are usually ions—atoms or combinations of atoms which have an excess of positive or negative charge.

Conduction in gases is in general an electronic process, but is non-ohmic. Electrons in a gas are more free to move than the electrons in a solid, and so an electric field can give rise to electrons of higher energy. These energetic electrons collide with gas atoms and knock off some of the more weakly bound atomic electrons. If the initial electron is fairly energetic it may give rise to a number of electron displacements, so that the number of free electrons increases rapidly. Some of these electrons will be recaptured, but Ohm's Law will not be obeyed. When a certain voltage is reached an electron avalanche will arise and extremely high currents can flow. This condition is known as gas breakdown.

This process of displacing electrons from atoms is one of the processes of ionisation, and it gives rise to charged bodies which will also move because of the electric field and contribute an ionic current. Being more massive than electrons, however, the ions will not move as rapidly, so that the majority of current carriers in a gas are electrons.

Insulators

The two last types of materials to be considered are insulators and semiconductors. The electrons in an atom have different energies, depending on the energy needed to remove each electron from the atom. When a number of atoms are collected together to form a solid the interactions between neighbouring atoms spread these electron energy levels into energy bands. The free electrons in the solid exist in these energy bands, which are separated by empty energy regions—forbidden zones in which there can be no free electrons. These bands and zones can be thought of as a pile of layers, one above the other, the greatest energies being in the upper levels. The uppermost band is the conduction band. Free electrons in this band can move through the solid under the effect of electric fields, and it is these electrons that contribute to the current through the material.

A metal or a conducting material has many free electrons in the conduction band, so that high currents are easily produced.

But in some materials this conduction band is devoid of free electrons, so that when a voltage is applied to such a material there are no free electrons to move through it and so no current can flow. Such materials are called insulators.

Semiconductors

A third class of materials exists which have a few energy levels just below the conduction band. The difference in energy between these levels and the conduction band is small enough for some electrons to receive enough energy by thermal agitation for them to rise into the conduction band. In course of time they fall back again to the lower levels, but at any time there are enough of these electrons in the conduction band to allow a small current to flow. If the temperature rises more electrons will be able to jump up into the conduction band, and a higher current can flow at higher temperatures. Compare this behaviour with that of metals.

Materials of this class, intermediate in behaviour between conductors and insulators, are called semiconductors. Considerable research has been carried out on these materials, and this has led to spectacular advances in the development of electronic devices such as transistors and integrated circuits.

The two most commonly used semiconductor materials are germanium and silicon, the atoms of both of which have four electrons in their outer, valence band. Both of these materials are crystalline, and in the crystal structure each of the four valence electrons form covalent bonds with the valence electrons of adjacent atoms: each atom is equidistant from its four adjacent atoms and each valence electron forms an electron pair, or covalent bond, with one from an adjacent atom. This condition exists at absolute zero temperature, when the atomic structure is complete and intact, each atom existing stably and at rest, and all covalent bonds being intact: semiconductor material is then a pure insulator. At normal temperatures, however, some of the valence electrons are given sufficient energy to break free from their covalent bonds and are thus available to act as negative current carriers. In breaking free from their covalent bonds, gaps, called *holes*, are created in the crystal structure of the semiconductor material. These holes act as positive current carriers since they attract electrons from neighbouring atoms and in doing so the hole reappears at the point in the crystal structure from which the electron has been gained. Thus at normal temperatures there is in semiconductor material a continuous generation of holes and free electrons, and if a potential difference is applied across a piece of the material the holes will move towards the negative terminal and the free electrons will move towards the positive terminal. As temperature increases so does the generation of holes and free electrons, so that the resistance of semiconductor materials falls with increase in temperature (compare this with the fact that the resistance of a conductor increases slightly with increase in temperature). It is frequently necessary with semiconductor devices to take steps to compensate for the effects of temperature variation.

n- and p-type Semiconductor Material

A preponderance of holes or free electrons in semiconductor material is established by introducing into the material—*doping* the material as it is called—small, controlled quantities of trivalent or pentavalent material. By introducing a trivalent material such as indium a preponderance of holes is established, since with each atom of indium that is introduced one valence electron of a germanium or silicon atom will be left unpaired, i.e. a hole in the semiconductor crystal structure is created. A trivalent impurity is termed an acceptor impurity, since the hole it creates will accept an electron from an adjoining atom, and as a hole represents a positive charge a semiconductor doped with a trivalent impurity is termed a p-type semiconductor. With a pentavalent substance such as antimony introduced into the semiconductor material an excess of free electrons is established, since the fifth valence electron of each pentavalent atom is left unpaired. Since a pentavalent impurity provides excess electrons it is termed a donor impurity, and as electrons carry a negative charge semiconductor material doped with a pentavalent impurity is termed an n-type semiconductor.

Semiconductor devices are based on what happens when a junction of p- and n-type semiconductor material is created—the junction must be a transition between n- and p-type

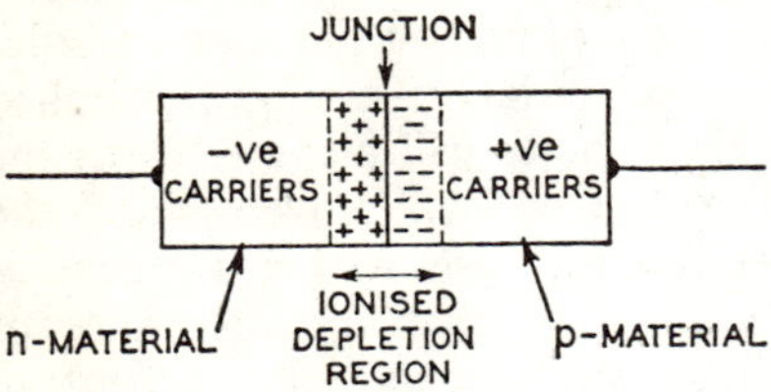

Fig. 5.—*Basic principle of a* ṗ ṅ *junction.*

regions within a single crystal structure. One way of achieving this result would be to change from acceptor to donor impurity doping during semiconductor material preparation. When the junction is first formed some of the electrons in the n-type semiconductor are attracted across the junction by the holes in the p-type material, and for the same reason holes will appear across the junction on the n side. The action of the pn junction is described in greater detail in Section 3.

THERMISTORS AND NON-LINEAR ELEMENTS

Both these circuit elements are semiconductors which introduce a non-linearity between some of the variables in the electrical circuit of which they form a part. The thermistor (temperature-sensitive resistor) exhibits a large variation in resistance with temperature variation. The non-linear element, one of the most popular examples of which is the voltage-dependent resistor, has a curvature in its current/voltage characteristic, i.e. as mentioned on page 12 the relation between current and voltage takes the form $I = kV^n$.

Thermistors have a large negative temperature coefficient, i.e. their resistance decreases rapidly with rise in temperature.

At room temperature, an increase in temperature of some common types of thermistor of approximately 20°C halves the value of resistance. The temperature coefficient decreases as the temperature rises, so that at 250°C an increase in temperature of approximately 50°C is required to halve the resistance value.

In the case of non-linear elements doubling the voltage applied across them may increase their current consumption about twenty times and trebling it about one hundred times.

2

THERMIONIC DEVICES

EMISSION OF ELECTRONS

It is now time to consider how to produce free electrons, both in the quantity required and in the required place.

Most solid materials, particularly metals, contain a large number of free electrons which are not attached to any particular atom but which can wander through the material. When these electrons approach the surface of the material they are pulled back by the effective charge on the rest of the material. But if an electron receives sufficient energy so that its speed as it approaches the surface is able to overcome these restraining forces, then it can escape and become a free electron. Methods of producing electrons inside the materials with enough energy to escape lead to ways of producing free electrons outside the material.

Thermionic Emission

When a material is heated up the increase in kinetic energy of the atoms and molecules is shared among the atoms and electrons. This energy is constantly being interchanged by reason of the frequent collisions between particles which occur in the material. These collisions may give some of the free electrons enough energy to escape: in a sense, the electrons are boiled off.

This method is of great importance in valves, for it is the most commonly used way of producing electrons in these devices.

Secondary Emission

When energetic external electrons strike the surface of a material they may collide with electrons just inside the material and transfer enough energy to the internal electrons to enable them to escape. As the energy of the incident electrons rises, the number of secondary electrons emitted will increase. However, when the energy of the primary electrons reaches a certain point the ratio of secondary to primary electrons is reduced. This is because the primary electrons are penetrating more deeply into the material and many of the collisions and interchanges of energy are occurring at such a depth that the extra energy gained by the internal electrons is lost by the time they get near the surface.

Electron photomultiplier tubes are an example of a device which makes use of photo-electric emission and secondary emission. Incident light falls on a piece of material which ejects some photo-electrons. A high voltage draws these electrons on to an electrode from which secondary emission of electrons occurs. The secondary electrons pass to another electrode at which more secondary electrons are produced, and so on, from electrode to electrode, the number of electrons being increased at each step. The number of electrons which reach the final electrode of the system may be as much as 10^8 times greater than the original number of photo-electrons, so that the tube can obviously act as a detector of very weak incident illumination.

Field Emission

The attractive force exerted by a very strong electric field may be sufficient to draw electrons out from the body of the material, and this is called field emission.

Radioactive Decay and Other Nuclear Reactions

All the methods of producing electrons so far considered have depended on the extraction of free electrons from inside the material. Another important way in which free electrons are produced is the natural process of radioactive decay of an atom. In this case the electrons which are liberated do sometimes come from the electron cloud which surrounds the nucleus, but more frequently they come from the nucleus itself.

Some of the neutral particles (neutrons) in the nucleus split up in some way, each neutron giving rise to two particles of equal and opposite charge. The positively charged particle is a proton, which, together with the neutron, is already known to exist in the nucleus: the negative particle is an electron, usually of high energy, which is then ejected from the new atom which has been formed. Electrons which are produced in this way are called beta-particles or beta-rays.

DIODES

A diode is a device that will offer a high resistance to the flow of current in one direction and a low resistance to the flow of current in the other direction. Current in this context means "conventional current" not electron flow: this is important to understand since it affects the understanding of circuit

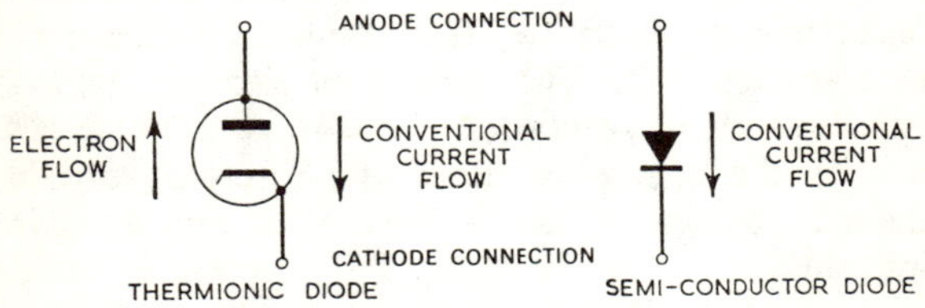

Fig. 1.—*Symbols for diodes.*

actions where the diode symbol is used. This symbol, according to British Standard Specification BS. 530: 1948 (as shown in Fig. 1), should point in the direction of flow of conventional current: the actual electron flow is in the opposite direction.

The diode has many uses in electronics. The main ones are rectification and detection, though clamping, clipping and sampling, or gate circuits, are also important. Diodes performing these functions are found in radar, television, control engineering, automation control and computers, to mention only a few of the branches of the wide field of electronics.

Types of Diode

There are a number of types of diode. Amongst the first types to be used were the copper-oxide rectifier and the selenium rectifier. For many years the thermionic diode tended to displace the use of these older types, and solid state rectifiers became mainly relegated to applications like power rectification. The pendulum has now swung the other way and solid state or semiconductor materials, like germanium and silicon, have displaced the thermionic diode from many of its applications because of their better characteristics and greater reliability, because they are smaller, do not need heater supplies, have a lower self-capacitance and operate over a wider frequency range. The thermionic diode is still in some ways superior to the semiconductor diode. On such points as reverse breakdown voltage, more rapid response to sudden voltage changes and an independence of ambient temperature changes it is preferred by engineers, but very rarely do these advantages become of sufficient importance to *demand* the use of a thermionic diode.

Forward-to-back ratio is a way of describing the resistance of a diode in the forward and the reverse direction. Thermionic diodes have a ratio of about 200 or 300 ohms to about 10^8 ohms. The semiconductor diode is better than this in the forward direction; in some cases it can be as low as a few ohms. In the reverse direction it can be as high as the thermionic diode although it is usually rather less.

Thermionic Diodes

The two-electrode valve or diode is often regarded as the parent of all thermionic valves and therefore an understanding of its function is important. It normally consists of a tungsten filament which is heated by an electric current. This heated filament raises the temperature of an oxide-coated nickel tube, called the cathode, around it. The cathode emits a cloud of "boiled-off" electrons which remain around the cathode to form the electron space charge. A second, larger nickel tube, called the anode, is placed around the cathode but not in contact with it. If the anode is made positive with respect to the cathode by a voltage, V_a, as shown in Fig. 2, then a current, I_a, will flow through the valve. This current is formed by the movement of electrons from the space charge to the anode. (Conventional current is, however, regarded as flowing from the anode to cathode—an important point to remember when reading circuitry.) Across the valve, due to its construction, there is a shunt capacitance, C_s, which can be a disadvantage under some circumstances.

Fig. 3 shows a typical thermionic diode of the miniature type—the Mullard EA76. It has effectively no reverse current

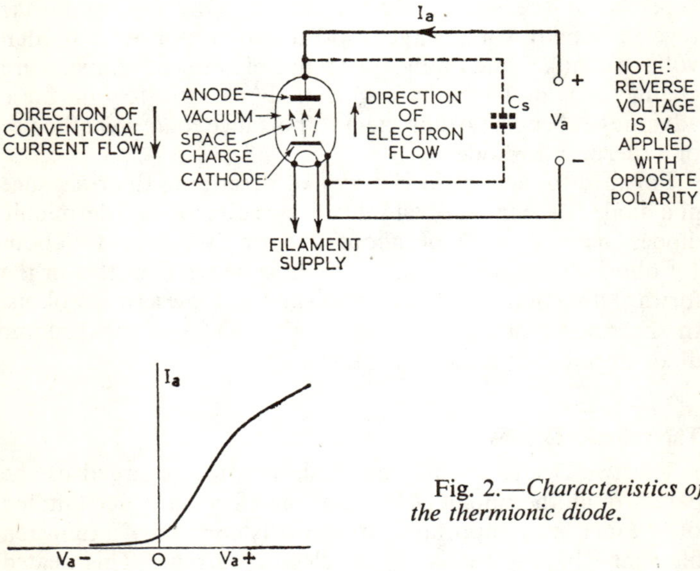

Fig. 2.—*Characteristics of the thermionic diode.*

and so the back resistance can be considered in excess of 10^9 ohms. The forward current is up to 9 mA and the diode will withstand a reverse voltage of up to 420 V. It is indirectly heated and requires a 6·3 V a.c. heater supply. It has an oxide-layer cathode and the anode is a nickel tube.

The advantages of a thermionic diode are as follows: (1) high reverse breakdown voltage, (2) fast current response to voltage changes across it, and (3) its characteristics are independent of ambient temperature changes. The disadvantages are: (1) forward resistance is relatively high, (2) high shunt capacitance, and (3) requires heater power supply.

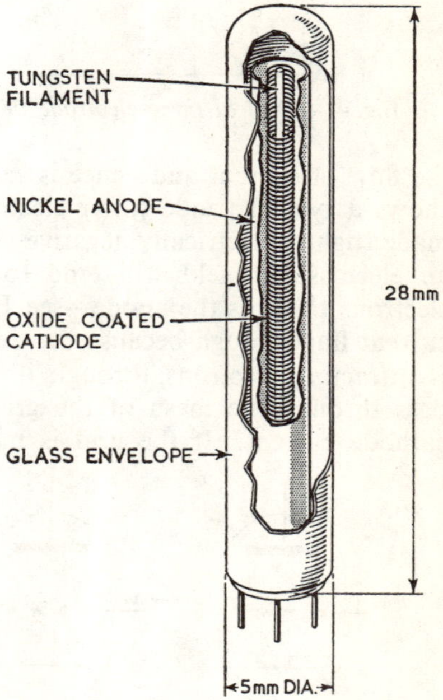

Fig. 3.—*Typical miniature thermionic diode, Mullard type EA76. Reverse voltage before breakdown*, 420 V. *Maximum forward current*, 9 mA. *Reverse current*, virtually nil. *Filament* ("*heater*") *voltage*, 6·3 V.

THERMIONIC TRIODES

In the diode electrons flow from the hot cathode towards the positive anode under the influence of the anode–cathode electrostatic field. The diode is a uni-directional device and "conventional current" can only flow from anode to cathode. A wire mesh inserted between the anode and cathode—physically nearest to the cathode—has a modifying effect on

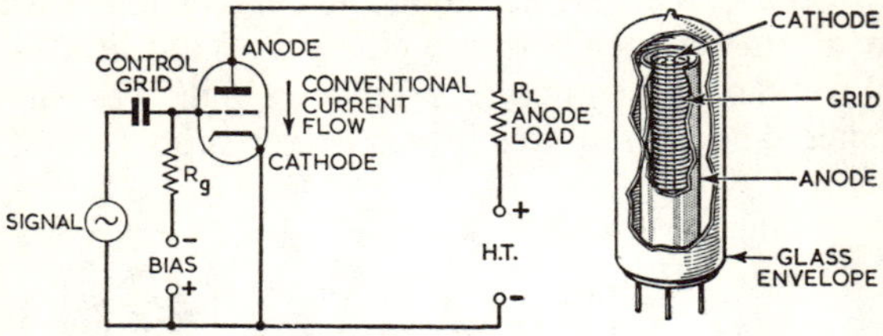

Fig. 4.—*Typical three-electrode or triode thermionic valve.*

the flow of current and hence is called a control-grid. Fig. 4 shows a typical triode (three-electrode) valve. If the grid is made slightly electrically negative with respect to the cathode, an electrostatic field will tend to retard the movement of electrons towards the anode—see Fig. 5. There will be some current flow though because the field from anode to cathode is attracting electrons through it. These escaping electrons pass through the mesh of the grid to form the net anode–cathode current. If the grid is made still more negative, a

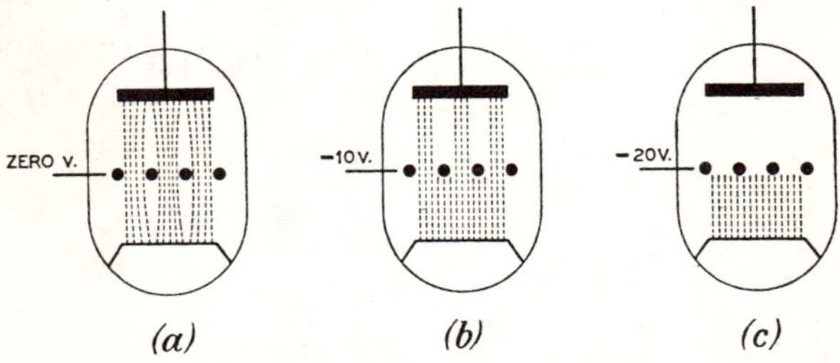

Fig. 5.—*Effect of grid voltage on electron flow in a triode valve.* (*a*) *No bias.* (*b*) *Working bias.* (*c*) *High negative bias* (*cut off*).

point will be reached at which the anode flow ceases: the valve is then in a cut-off condition.

Characteristics

Since the grid is nearer to the cathode than is the anode, it has more influence per volt than the anode. This means that if the grid is made one volt more (or less) negative, it will need a great deal more than one volt increase in anode voltage to maintain the anode current at the same amplitude. If, say, 50 volts change on the anode were needed to compensate for 1 volt change on the grid, then the valve is said to have an amplification factor (μ) of 50.

The mutual conductance of a valve, g_m, is the ratio of the change in anode current for a change in grid voltage if the anode volts are not altered. This figure is expressed in milliamperes per volt and indicates the "goodness" of a thermionic valve.

Fig. 6 shows a graph of the anode current against grid voltage. A thermionic valve is normally biased negatively and the grid is not allowed to be positive except in special cases. If the grid becomes positive it may collect too many of the electrons and its characteristics alter producing distortion. The input signal is normally applied between the grid and cathode as shown in Fig. 4 so that even on the positive half-cycles of the input voltage the control grid is always a fraction of a volt

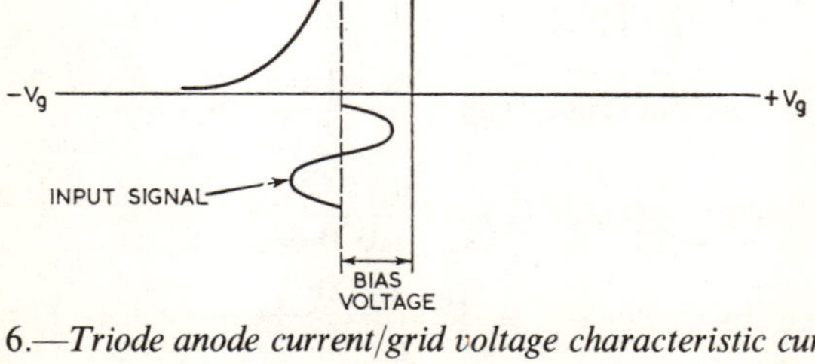

Fig. 6.—*Triode anode current/grid voltage characteristic curve.*

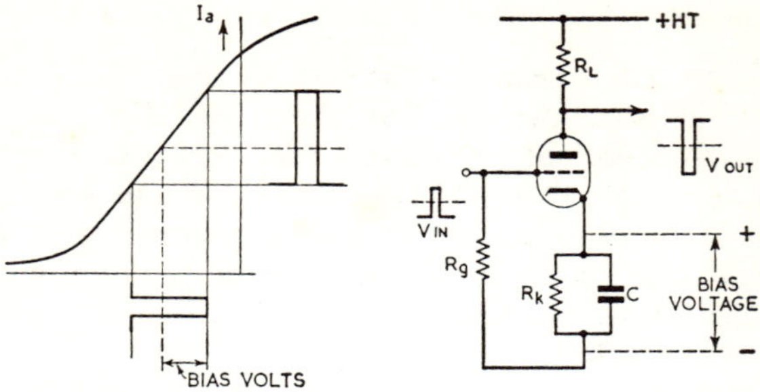

Fig. 7.—*Biasing a valve used as an amplifier so that it works on the linear part of its* I_a/V_g *characteristic: class* A *operation of a thermionic triode valve. Note the inversion of polarity of the waveform at the anode.*

negative. The negative half-cycles, of course, will make it a few volts negative. In this way the anode current is varied in step with the grid voltage. This current flowing through a resistor in the anode circuit will produce across it a voltage which is an amplified replica of the input voltage. With some valves this amplification or gain can be a 100 times. How much it is depends upon the amplification factor (μ), the anode resistance (r_a) of the valve and the anode-load resistance R_L. Gain may be expressed as follows:

$$G = \mu \,.\, R_L/r_a + R_L.$$

This type of linear amplification is called class A amplification. The valve is always working on the linear, i.e. the straight, part of the I_a/V_g characteristic, as seen in Fig. 7. The subject of amplification is taken further in Section 5.

The following relationships exist between the main characteristics of a triode valve:

$$\text{Mutual conductance } (g_m) = \frac{\mu \times 1{,}000}{r_a} \text{ mA/volt,}$$

$$\text{Amplification factor } (\mu) = \frac{g_m \times r_a}{1{,}000},$$

$$\text{Anode resistance } (r_a) = \frac{\mu \times 1{,}000}{g_m} \text{ ohms.}$$

While mutual conductance is a direct indication of the goodness of a valve under stated conditions, amplification factor means little unless resistance is taken into account. There may, for example, be no advantage in doubling the amplification factor, unless the circuit conditions are such that a valve of higher anode resistance can be used to greater advantage. Valves with an amplification above about 30 are generally known as high-μ types.

American manufacturers use the term transconductance instead of mutual conductance: this gives the same indication of valve goodness, but is derived from the ability of the electron stream to conduct, rather than—as in the case of mutual conductance—to oppose, and is often given not in mA/volt but in micromhos.* (1 mA/volt equals 1,000 micromhos.)

Transconductance is obtained by dividing the change of anode current by the change of grid voltage.

Limitations of Triode Amplification

The amplification factor of a triode valve is governed by the distance of the grid from the cathode, and by the spacing of the grid wires: a close-mesh grid exercises greater control over the electrons than one of coarser construction having wires spaced relatively far apart. Unfortunately, however, as the mesh of the grid is increased, certain difficulties arise.

Firstly, the area of the grid electrode will increase, and thus increase the capacitances between the electrodes. The problem of inter-electrode capacitances has been largely overcome with the introduction of miniature types in which the inter-electrode capacitances are as little as one or two picofarads.

Secondly, not all the electrons emitted by the cathode reach the anode, and these, together with any electrons repelled by a negatively charged grid, tend to accumulate in the space between the cathode and the grid, giving rise to what is known as the "space charge". The fresh electrons emitted by the cathode have to break through this charge in order to reach

* Mho, unit of conductance, is the reciprocal of resistance.

the anode. Since a closely wound grid exercises more control over the electron flow, it will tend to drive a greater number back, thus increasing the magnitude of the space charge and impeding the cathode–anode flow.

TETRODES AND PENTODES

The limitations of the triode can be overcome by inserting a further grid between the control grid and the anode; this grid when earthed, normally via a suitable capacitor (see C, Fig. 8), acts as an electrostatic shield, and thus greatly reduces the capacitative coupling between grid and anode. Furthermore, by connecting a positive potential on this electrode, it will assist in attracting electrons from the cathode, and thus help to overcome the space charge. By the time the electrons reach the second grid (screen grid), they are travelling sufficiently fast for most of them to pass through and reach the anode. By in this way reducing the limitations of the triode valve, the

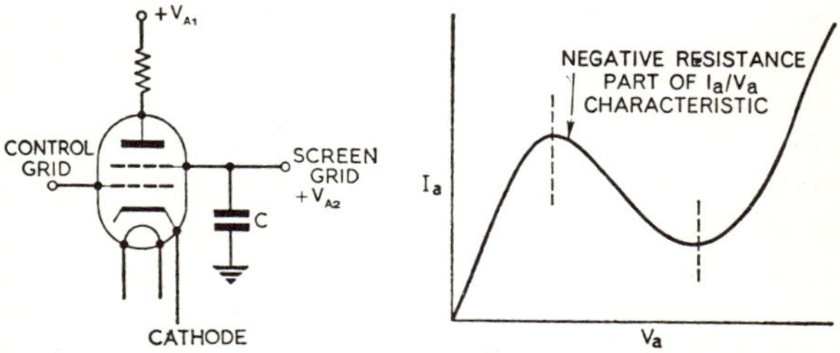

Fig. 8.—*The tetrode valve.*

control grid can be made of much finer mesh and the amplification factor raised to 1,000 or more.

A schematic diagram of a tetrode valve, showing connections to the electrodes, together with typical I_a/V_a characteristic curve, are shown in Fig. 8.

The characteristic shows that over a portion of the curve as the anode voltage is increased the anode current falls—provided that the screen voltage is kept constant. This kink in the curve —the portion of negative resistance—is caused by the fact that at a certain anode voltage the electrons strike the anode with such force that other electrons are displaced from the anode; that is, secondary emission occurs. These electrons will fall back into the inter-electrode space and, if the screen grid voltage is higher than the anode voltage, will be attracted to the screen grid. As the anode voltage is increased above the screen voltage, the anode current will begin to rise again until a saturation value is reached.

The negative resistance part of the tetrode curve is from the point of view of amplification obviously a disadvantage. It has, however, a practical value in that it can be used to sustain an oscillator circuit. A tuned circuit will not oscillate indefinitely because its energy is dissipated in the resistance of the *L* and *C* components. This resistance can be regarded as equivalent to placing a large positive resistor across a lossless inductor and capacitor. If a negative resistance of the same value as this positive resistor is placed across the circuit, it will cancel the positive resistance so that oscillations will continue indefinitely. The tetrode valve can be used to do this. Many tetrodes have beam-forming plates between the screen grid and anode and are in many ways akin to pentodes.

Pentodes

The kink in the I_a/V_a characteristic of the tetrode valve can be largely eliminated by inserting a further grid between

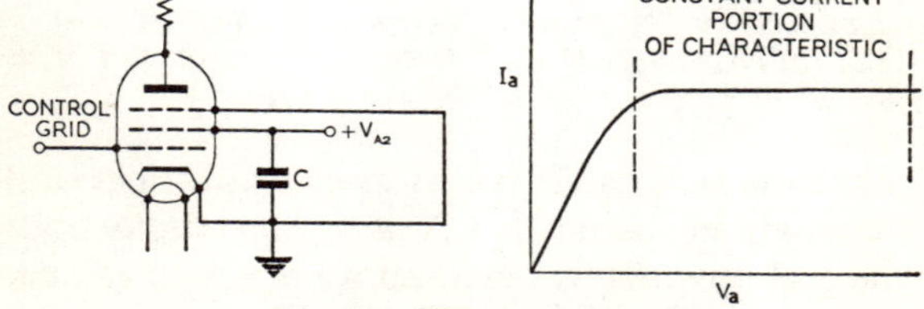

Fig. 9.—*The pentode valve.*

the screen grid and the anode. This electrode, known as the suppressor grid, is, for most applications, connected to the cathode. It is thus negative with respect to anode, and will repel the electrons dislodged from the anode by secondary emission.

A typical pentode I_a/V_a characteristic is shown in Fig. 9. It can be seen that over the major portion of the characteristic the current is largely independent of anode voltage changes. The pentode can, therefore, be looked upon as a constant current source. The amplification of a pentode valve can be approximated to $g_m \cdot R_L$. One disadvantage of the pentode valve is that, due to the number of electrodes, more "noise" is generated in it than in a triode.

THYRATRONS

The thyratron is a triode or tetrode valve the electrodes of which are contained in a bulb partially filled with an inert gas such as helium or neon (cf. a valve where the envelope is as closely as possible evacuated or *de-gassed*). The second grid of a tetrode thyratron is called the shield grid. The gas-filled

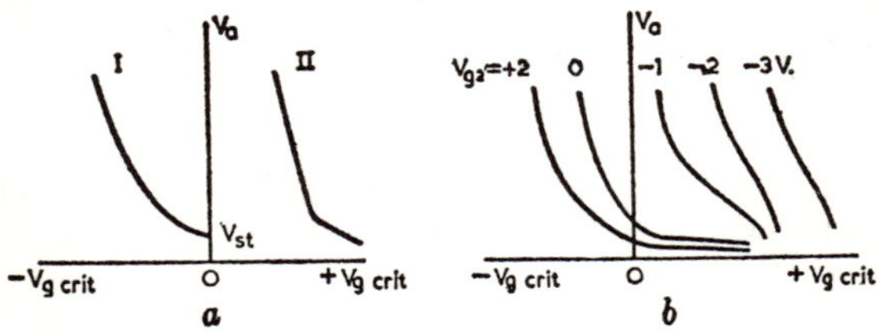

Fig. 10.—*(a) Typical control characteristics of thyratrons.* I *negative control characteristic;* II *positive control characteristic.* (*b*) *Showing how the control characteristics of a tetrode thyratron can be modified by varying the voltage applied to the shield grid.*

valve is known as a "soft" valve in contradistinction to the "hard" vacuum-enclosed valve. A thyratron is non-conducting unless the grid potential is raised above a certain value known as the critical grid voltage and the anode voltage is above a value known as the firing, striking or arcing value. When the grid voltage is raised above the critical value and the anode voltage is above the striking value a small current flows and, by reason of the ionisation of the internal gas, the anode current rises rapidly to a maximum value which is governed by the anode voltage and the resistance in the anode circuit. When the tube is conducting no further control of the anode current can be achieved, and the anode current will continue to flow until the control grid voltage is reduced below the critical value and the anode voltage is reduced below the striking value.

There is a definite relationship between the critical value of grid voltage and the anode striking voltage. Typical control characteristics are shown in Fig. 10. It will be seen that thyratrons are available having either negative or positive control characteristics. The anode striking voltage can be varied by altering the bias applied to the grid. For a thyratron with a negative control characteristic, the more negative the potential applied to the grid the higher the anode voltage required to make the valve strike. If, for instance, with -1 V negative bias applied to the grid the striking voltage is 80 V,

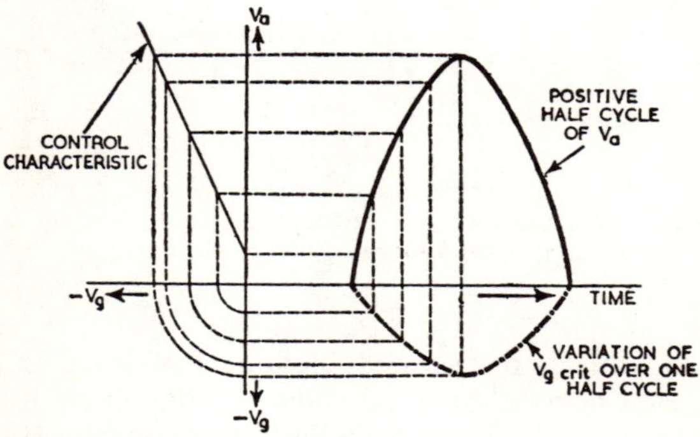

Fig. 11.—*Change of critical grid voltage over a positive half-cycle of anode voltage.*

changing the negative grid bias to -2 V will result in the striking voltage being raised to about 100 V. The ratio between the increase in grid bias and the increase in the striking voltage is called the *control ratio* of the thyratron. Fig. 10 (b) shows how the control characteristics of a tetrode thyratron can be changed by varying the potential at the shield grid.

When a thyratron is operated with an a.c. anode supply, it acts as a half-wave rectifier, passing current only during the positive half-cycles (provided that the grid voltage is above the critical value). As the critical value of the grid voltage at any instant depends on the instantaneous value of the anode voltage, the critical grid voltage is not constant but varies throughout the cycle. This is indicated in Fig. 11. By adjusting the steady value of d.c. grid voltage (Fig. 12) the proportion of each positive half-cycle during which anode current will flow can be regulated and thus the mean value of anode current controlled. Alternatively, the proportion of each positive half-cycle during which the thyratron conducts can be controlled by applying a combination of d.c. and a.c. grid voltage as

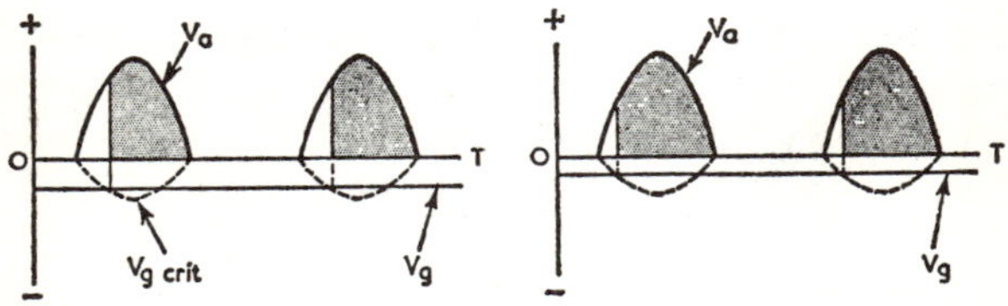

Fig. 12.—*When a thyratron is operated with an a.c. anode supply voltage the proportion of the positive half-cycle during which it conducts can be controlled by varying the grid voltage.*

shown in Fig. 13, or the d.c. component of the grid voltage may be kept below the critical value and the phase of the a.c. component varied as shown in Fig. 14. Control can also be effected by applying positive-going pulses instead of sinusoidal voltages to the grid.

Thyratrons can be used as high-speed electronic switches and to control comparatively heavy currents, avoiding the heavy

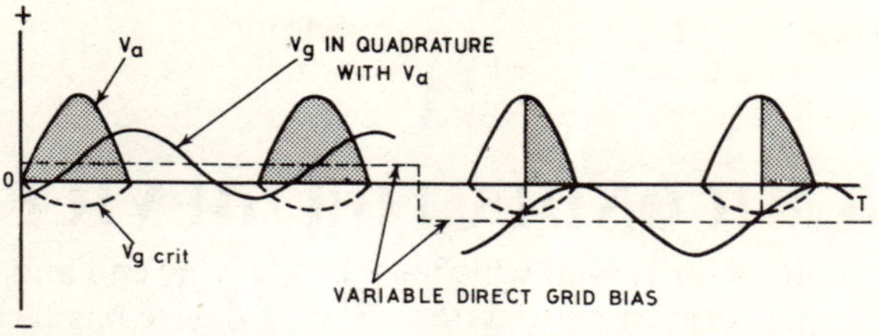

Fig. 13.—*If the grid voltage applied to a thyratron consists of a constant a.c. component in quadrature with (i.e. with 90° phase difference between) the a.c. anode voltage and an adjustable d.c. component, variation of the d.c. component will determine the proportion of each positive half-cycle of anode voltage during which the thyratron conducts, and hence the mean value of the anode current. This is termed vertical shift control.*

destructive sparking which accompanies the operation of mechanical switches or contactors. The properties of the thyratron can also be used, in conjunction with a capacitor to

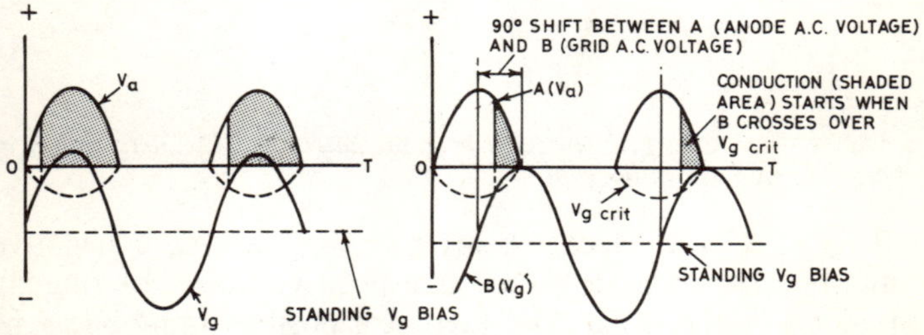

Fig. 14.—*Horizontal shift control of a thyratron. The grid voltage consists of a standing d.c. bias on which is superimposed an a.c. voltage of the same frequency as the a.c. anode voltage. By shifting the phase of the a.c. component, the proportion of each positive half-cycle of the a.c. anode supply during which the thyratron conducts can be controlled.*

hold and quickly release an electric charge and pulse control of the grid, as a sawtooth oscillator (see Section 6). The semiconductor equivalent to the thyratron is the silicon-controlled rectifier (see Section 3).

3

SEMICONDUCTOR DEVICES

A semiconductor crystal which is p type at one end and n type at the other forms a pn junction. Such a device has rectifying properties.

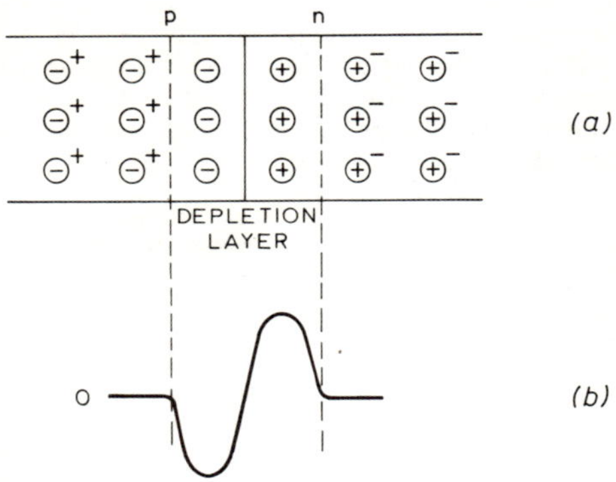

Fig. 1.—(*a*) *Charged particles in a* pn *junction.* (*b*) *Distribution of charge through a* pn *junction.*

In Fig. 1(a) an acceptor atom is depicted as a negative immobile ion (ringed) plus a free positive hole (not ringed). Similarly in the n material there are positive immobile ions and free electrons. When the junction is formed, diffusion of current carriers in the vicinity of the junction will cause holes and electrons to recombine leaving unneutralised negative ions in the p region and positive ions in the n region. This results in a thin depletion layer (i.e. a layer depleted of current carriers) being formed round the junction.

Although the net charge on the device is zero, examining the crystal throughout its length we find the charge distribution shown in Fig. 1(b); this charge difference across the depletion layer, which is now an insulator, defines a charged capacitance across which there will be a junction potential V_j. This potential limits the extent of the hole/electron recombination process because its polarity is such as to erect a potential energy barrier to both holes and electrons.

With no externally applied voltage the device is said to be *in equilibrium.* However, movement of current carriers still exists. In the first place, thermally generated hole/electron pairs will be formed in both regions and minority carriers in the vicinity of the junction will be swept across it by V_j. Secondly, since the distribution of thermal energy is uneven, some majority carriers will gain enough energy to cross the junction and cancel out the minority carrier current.

Reverse Bias Condition

An externally applied voltage V connected as shown in Fig. 2(a) increases the junction potential and reduces the

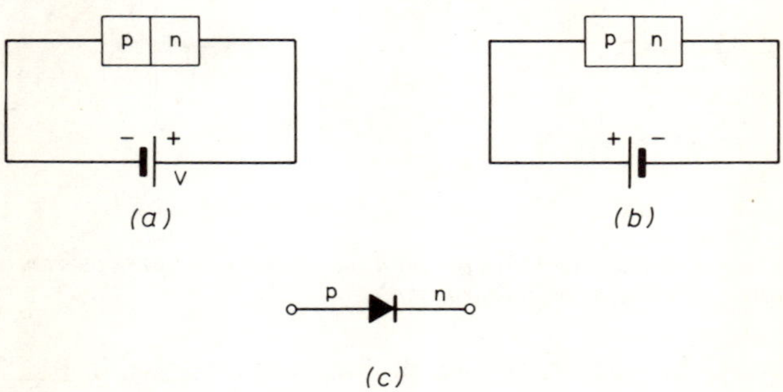

Fig. 2.—(*a*) *Reverse-biased* pn *diode.* (*b*) *Forward-biased* pn *diode.* (*c*) *Diode symbol.*

majority carrier flow. The minority carrier current is not affected and the net effect is a small reverse leakage current I_0, which does not increase with V unless the junction breaks down but which does increase with temperature. The movement of charge may be stated as follows: an electron from the p region crosses the junction into the n region leaving the p region positively charged and the n region negatively charged. This is compensated for by an electron flowing from the n region into the battery, and an electron flowing from the battery into the p region. A similar sequence can be drawn up beginning with a hole passing from the n region into the p region.

Forward Bias Condition

If the external voltage V is now connected as in Fig. 2(b) the junction voltage is lowered (not reversed) and majority carriers can more easily pass the reduced barrier potential. The forward current flow rises as V is increased to reduce further the junction voltage.

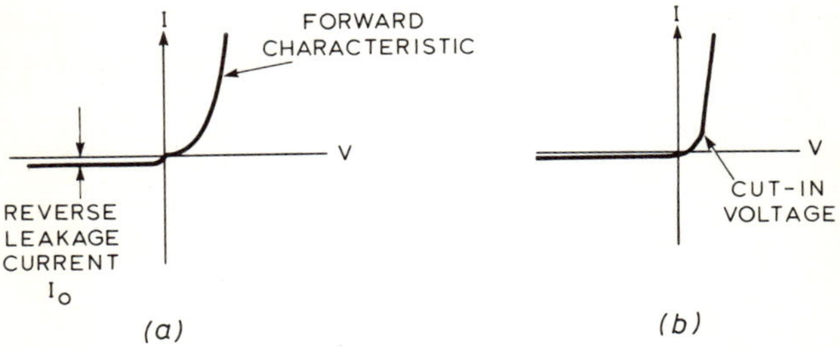

Fig. 3.—I/V *characteristic for* pn *diode: (a) with reverse current scale magnified; (b) drawn on uniform scale.*

The I/V characteristic of a pn junction is shown in Fig. 3(a). It is similar to that of a thermionic diode, the main differences being: (a) there is no reverse leakage current in a thermionic diode; (b) the pn characteristic passes through the origin whereas the thermionic diode has a small forward current flow at $V = 0$ V. The forward and reverse current scales of Fig. 3(a) are different, otherwise it would be difficult to see the nature of the reverse characteristic. Drawing the characteristic on a common current scale as in Fig. 3(b) we find a region where, under forward bias conditions, the current is so small that the device does not appear to conduct at all. The voltage at which forward conduction begins to increase sharply is called the *cut-in voltage* and is about 0·2 V for germanium and 0·6 V for silicon.

JUNCTION (BIPOLAR) TRANSISTOR

The junction transistor is a three-layer device which, in essence, consists of two pn junctions connected back to back.

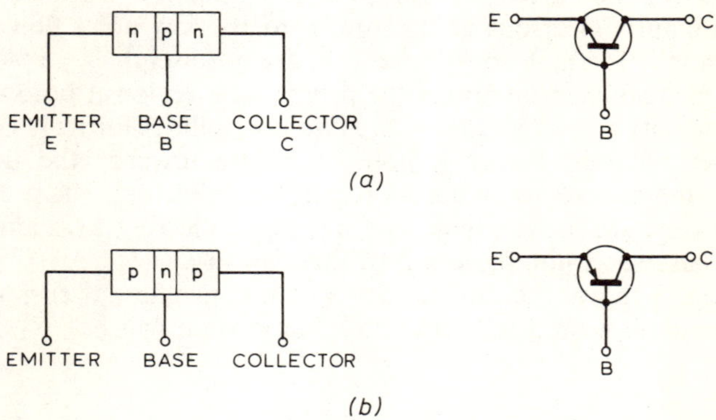

Fig. 4.—(*a*) npn *and* (*b*) pnp *transistors*.

It may be npn or pnp in construction, as shown in Fig. 4. The three regions are called emitter (E), base (B) and collector (C). Two important features of its construction are that the base region must be thin and the doping of the emitter region must be much more concentrated than that of the base. The alternative description—bipolar transistor—implies that its operation depends on the flow of majority/minority carriers.

Transistor Action

To explain the action of the npn transistor we should first observe the biasing arrangements of Fig. 5. Here the device

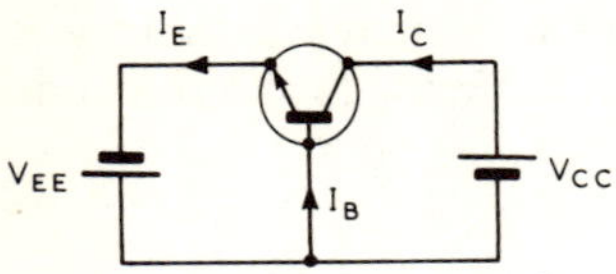

Fig. 5.—*Biasing and actual current directions in an* npn *transistor.*

is connected in its common base (CB) mode, i.e. the base lead is common to both d.c. supplies. The E–B junction is forward biased and B–C is reverse biased. Since the emitter doping is so heavy the current crossing the E–B junction will be predominantly electrons going from E to B. Any holes flowing from B to E which contribute to I_E are not useful.

Once an electron enters the p type base region it becomes a minority carrier and there are two possibilities for it; it may either recombine with a hole or diffuse towards the B–C junction and be swept across it into the collector. The useful electrons are those which pass into the collector; by making the base very thin, these will be the majority.

Fig. 5 shows the direction in which conventional currents actually flow, and the Kirchhoff relationship will be:

$$I_E = I_C + I_B \tag{1}$$

Since I_C is the output or useful part of I_E, I_B may be regarded as the wasteful part and consists of the two components mentioned, i.e. holes flowing from B to E and electrons lost in recombination in the base region. The same explanation of transistor action will serve for a pnp transistor if current directions and bias supplies are reversed and the words hole and electron interchanged.

It is usual practice in network theory to show all currents flowing inwards as in Fig. 6. Equation (1) must now be re-written as:

$$I_E + I_B + I_C = 0 \tag{2}$$

Fig. 6.—*Normal network current directions.*

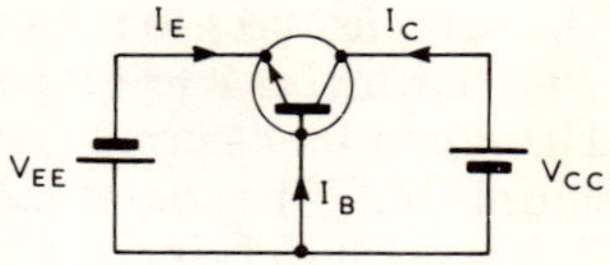

For an npn transistor I_E will now be negative. To relate input current I_E and output current I_C we have

$$I_C = -I_E - I_B$$
$$= h_{FB} I_E$$

where h_{FB} is called the common base current gain. Since h_{FB} must be less than 1 it is a loss and, using the inward flowing current convention, it will also be negative, $-0{\cdot}99$ being a typical value.

Common-base Characteristics

Typical output characteristics for a CB çonnected transitor, i.e. I_C vs V_{CB} for a number of values of I_E, are shown in Fig. 7.

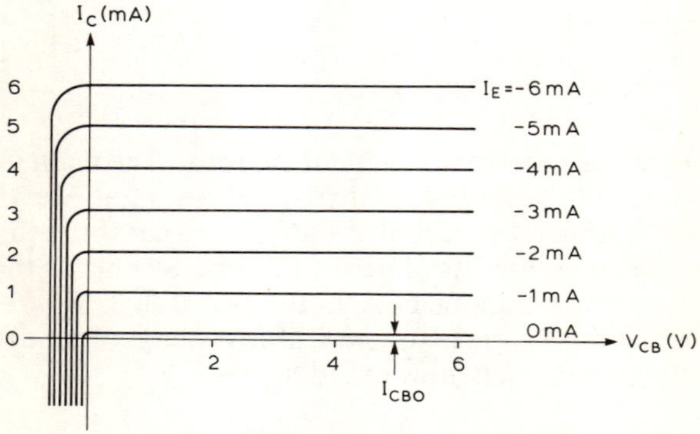

Fig. 7.—*Common-base collector characteristics.*

The main features to notice are:

(a) Each characteristic is almost a horizontal straight line. These constant current characteristics imply a high output impedance. The constant current is maintained at $V_{CB} = 0$ V. This is because I_C is created by minority carriers from the base crossing into the collector so an external bias is not essential.

(b) I_C is almost completely dependent on I_E—the device is said to be current controlled—and virtually independent of V_{CB}.

(c) For negative values of V_{CB} the C–B junction becomes forward biased, so a large current flows from C to B.

(d) When $I_E = 0$ V there is a small leakage current I_{CBO} passing from C to B consisting of thermally generated current carriers.

Common-emitter (CE) Configuration

A simple reconnection of the bias supplies to produce the common emitter circuit of Fig. 8 considerably affects the

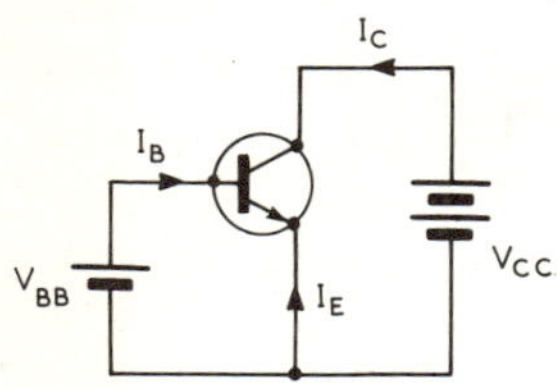

Fig. 8.—*Common-emitter circuit.*

operation of the device as a circuit element. The main feature to notice is that the input current is now I_B so we should expect a considerable current gain I_C/I_B. That this will be so can be seen in a number of ways. For one, as long as the bias conditions are maintained (i.e. E–B forward, B–C reverse) the transistor is not aware, so to speak, of any change of connection. Then the current gain, now called h_{FE}, is

$$h_{FE} = \frac{I_C}{I_B} = \frac{I_C}{-I_C - I_E}$$

$$= -\frac{I_C/I_E}{(I_C/I_E)+1}$$

$$= -\frac{h_{FB}}{h_{FB}+1}$$

Using the value for h_{FB} already given,

$$h_{FE} = -\frac{-0{\cdot}99}{-0{\cdot}99+1} = +99$$

A more physical explanation would be to consider varying the voltage V_{BE}. Although a transistor is normally regarded as a current-controlled device (i.e. the output current I_C is proportional to the input current I_B), nevertheless we can see that increasing V_{BE} increases the current flowing across the E–B junction and transistor action makes most of this into collector current. The base current produced in this operation is small.

Common-emitter Characteristics

The CE output characteristics, i.e. I_C vs V_{CE} for a number of values of I_B, are shown in Fig. 9. By contrast with the CB

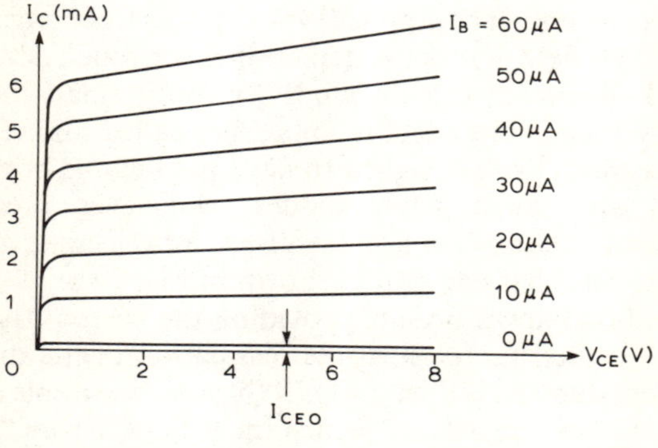

Fig. 9.—*Common-emitter collector characteristics.*

characteristics they are greater in slope and they do pass through the origin. V_{CE} is divided between the two junctions, and when it falls to a value less than the forward bias on the E–B junction the direction of V_{BC} reverses and I_C falls.

With the base open circuit $I_B = 0$ and the leakage current flowing through the device is I_{CEO}, which is greater than I_{CBO}. This is because V_{CC} forward biases the E–B junction, so the leakage current will mainly originate as majority carriers in the emitter rather than thermally generated minority carriers. If it is required to reduce the leakage current to I_{CBO}, V_{BE} must be made negative.

ZENER DIODES

Since the depletion layer in a pn junction is devoid of current carriers, it forms a region of high resistivity and most of the applied voltage will be dropped across it. The reverse voltage may be increased to such an extent that the diode will break down. There are two mechanisms which can cause this.

(1) Avalanche—high velocity minority carriers travelling through the depletion layer dislodge valence electrons. The effect is cumulative.

(2) Zener effect—some devices are made with very narrow depletion layers. The applied voltage therefore produces a large electric field across the depletion layer which is sufficient to break down co-valent bonds for only small values of applied voltage. The effect is sometimes called tunnelling.

Devices which are designed to have predictable breakdown voltages are called zener diodes, whichever breakdown mechanism occurs, and are useful for establishing reference voltages. The characteristic is shown in Fig. 10(a).

When breakdown occurs, providing the current that flows does not exceed the permissible dissipation of the diode, no permanent damage is done. Fig. 10(b) shows a simple application as a voltage regulator. Whilst the voltage across the zener diode is held constant, correct operation can only be achieved when a series resistor R is available to drop the voltage from its supply level to the output required. Two zener diodes connected back-to-back can be used for a.c. regulation, producing a clipped a.c. waveform.

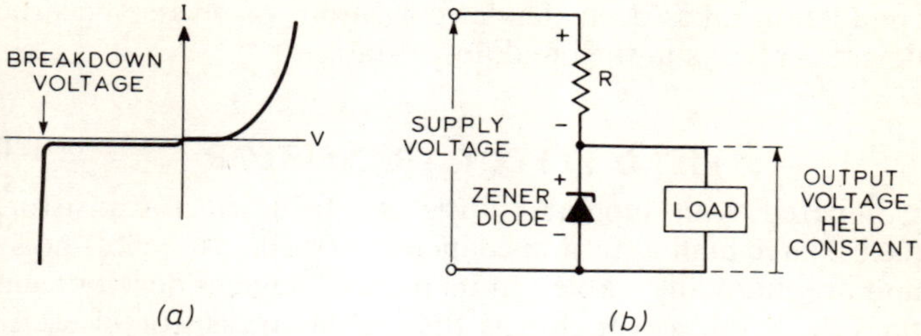

Fig. 10.—(*a*) *Zener diode characteristic*. (*b*) *Simple application of a zener diode as a voltage regulator*.

TUNNEL (ESAKI) DIODE

The tunnel diode, sometimes called the Esaki diode after its inventor, is a pn junction in which the doping is very heavy. This results in a narrow depletion layer in a state of breakdown even without any external bias at all. As forward bias is applied conduction will be heavy, as shown between points 1 and 2 on the characteristic of Fig. 11. Increasing *V* reduces the

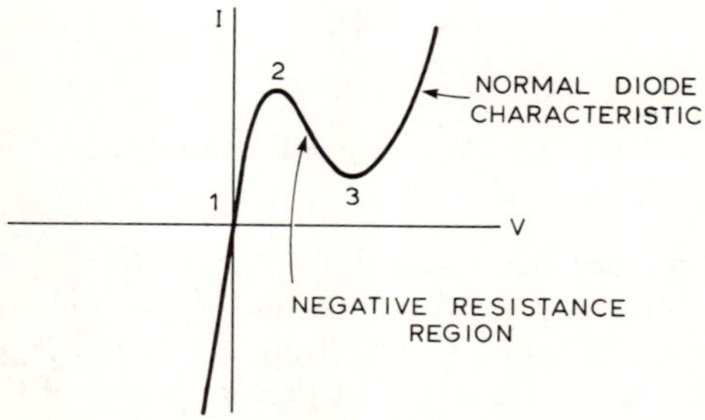

Fig. 11.—*Tunnel diode characteristic*.

junction potential and the device comes out of breakdown and current I falls with increasing V. This is the device's most useful region, one of negative resistance, as indicated between points 2 and 3. From point 3 onwards normal forward conduction takes over. Under conditions of reverse bias the device is always in the breakdown state.

FIELD EFFECT TRANSISTOR

Compared with bipolar transistors, field effect transistors (f.e.t.s) have high input impedance ($>10^6\ \Omega$), can be less noisy and are thermally stable. In its present stage of development the f.e.t. is not as capable as the bipolar transistor of useful gain at very high operating frequencies. Nevertheless its advantages suggest that it will supersede the bipolar transistor for an increasing number of applications.

To understand the operation of an f.e.t. consider Fig. 12.

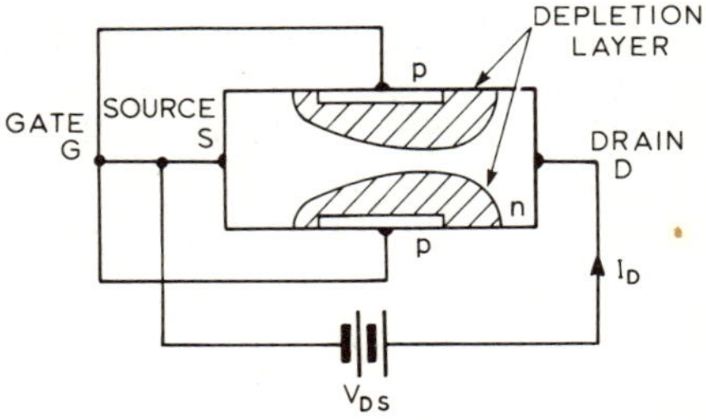

Fig. 12.—*Structure of n-channel f.e.t.*

The bar of n type silicon, called the *channel*, forms pn junctions with the p elements of the *gate*, and depletion layers occur around these junctions. Connection of a battery V_{DS} as shown thickens the depletion layers at the drain end of the channel because at this end the channel-to-gate diodes are reverse biased.

For small values of V_{DS} a current I_D proportional to V_{DS} flows through the channel consisting of electrons, i.e. majority carriers. However, when V_{DS} reaches a certain voltage, called *pinch-off*, V_P, the effective channel width has become so narrow due to the constriction caused by the depletion layer that the current ceases to increase between the increasing V_{DS}.

If a battery is connected to make V_{GS} negative the gate–channel junction will be reverse-biased and the depletion layers widened. This has the effect of reducing the pinch-off voltage. Typical characteristics are shown in Fig. 13 and are similar in shape to those of a pentode valve.

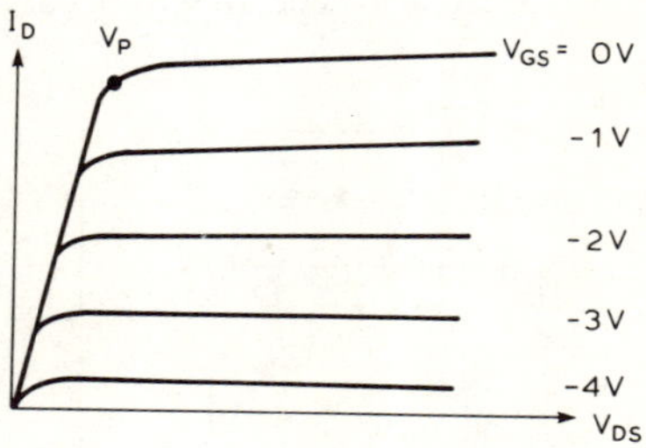

Fig. 13.—*Drain characteristics of f.e.t.*

The device described here is an n-channel f.e.t. The p-channel f.e.t. has an n-type gate and reversed voltage and current directions. The symbols are shown in Fig. 14.

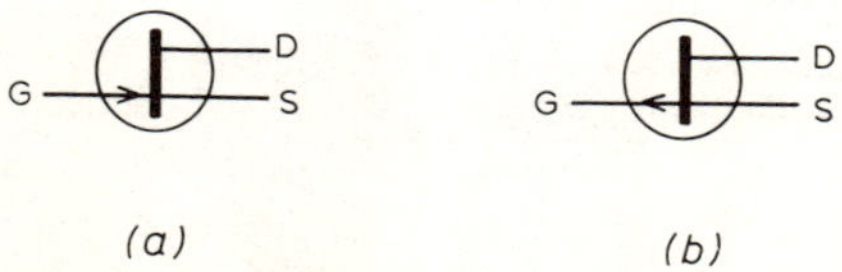

Fig. 14.—*F.E.T. symbols:* (*a*) *n-channel;* (*b*) *p-channel.*

The equivalent circuit, parameters and external circuitry used are similar to those of a triode valve, with the terminals corresponding as follows:

Source—Cathode
Drain —Anode
Gate —Grid

M.O.S.F.E.T.

Another type of f.e.t. is the *metal oxide semiconductor* f.e.t. or m.o.s.f.e.t. It is also known as the *insulated gate* f.e.t. or i.g.f.e.t. There are two different types of m.o.s.f.e.t., one of which, the *enhancement* type, is shown in Fig. 15(a). It consists of a p-type substrate (it could be n-type, in which case all subsequent

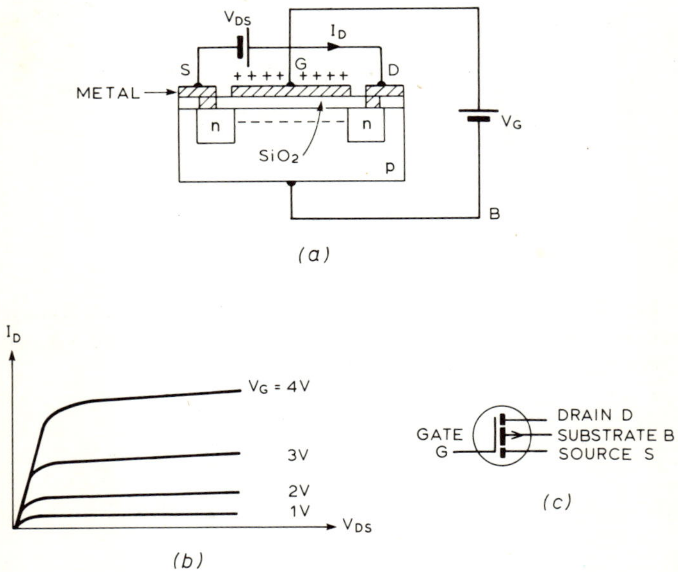

Fig. 15.—*p-channel enhancement-type m.o.s.f.e.t.* (*a*) *Structure.* (*b*) *Drain characteristics.* (*c*) *Symbol.*

polarities and dopings will be reversed), into which two heavily doped n regions are diffused. SiO_2 is grown over the surface. The two n regions are the *source* (S) and *drain* (D). Holes are cut into the glass to make external connections. The p substrate between source and drain forms the channel and a metal layer covers the SiO_2 opposite the channel to form the *gate* (G). The gate and channel make up the plate of a capacitor with the SiO_2 as its dielectric.

With the substrate earthed, a positive voltage on the gate will produce negative charges in the channel and increase the conductivity between source and drain. The drain characteristics are shown in Fig. 15(b), the symbol in Fig. 15(c).

The *depletion* type m.o.s.f.e.t. is shown in Fig. 16(a). The difference between this and the enhancement type is that here

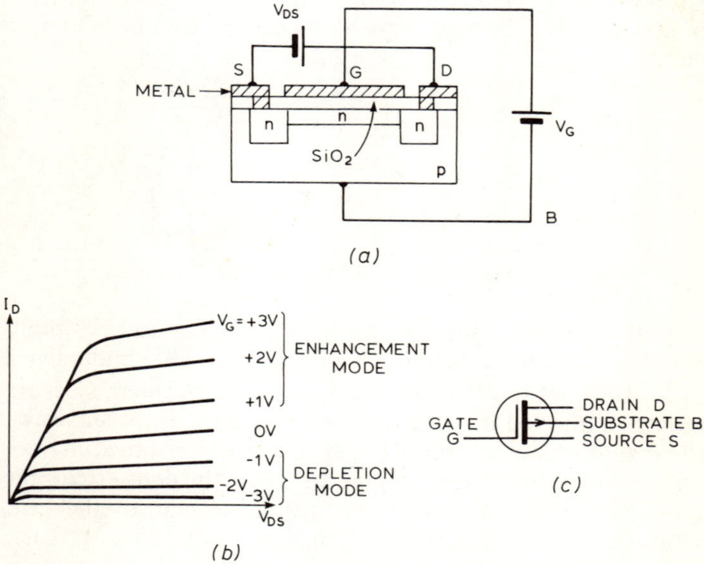

Fig. 16.—*p channel depletion-type m.o.s.f.e.t.* (*a*) *Structure*. (*b*) *Drain characteristics*. (*c*) *Symbol*.

the channel between the heavily doped n-type drain and source regions is a lightly doped n region. With this arrangement there is an appreciable drain current I_O even with zero voltage on the gate. The conductivity will be reduced if V_{GS} is made negative, and the device is said to be operated in its *depletion mode.* Alternatively, if V_{GS} is made positive, the device will be in its *enhancement mode* since the channel conductivity will be increased. The drain characteristics are shown in Fig. 16(b), the symbol in Fig. 16(c).

UNIJUNCTION TRANSISTOR

The unijunction transistor (u.j.t.), originally called the double-base diode, consists of an n-type bar with a contact at each end called Bases 1 and 2 (B1 and B2). About halfway along the bar there is a p region called the emitter (E). The arrangement is shown in Fig. 17 together with its symbol and normal bias circuit.

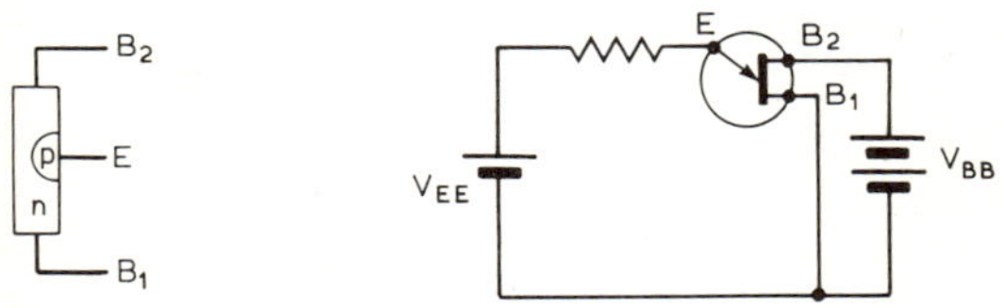

Fig. 17.—*Unijunction transistor and normal bias circuit.*

With E open-circuit the voltage $V_{EB1} = V_{EO}$ and is simply the result of the division of V_{BB} along the bar. The conductivity of the bar does not change until $V_{EE} > V_{EO}$. When this occurs holes are injected into the bar and V_{EB1} falls. In other words the region between E and B1 has a negative resistance characteristic, with the emitter voltage falling as the emitter current rises. The construction of the u.j.t. is similar to the f.e.t. However the u.j.t. emitter is smaller than the f.e.t. gate. Operationally, the u.j.t.'s useful mode is with the E–B1 junction forward biased, whereas the f.e.t.'s normal operation is with gate to source reversed.

THYRISTORS

Some confusion exists over the use of the term *thyristor*. Here it is taken as a generic title, the *silicon controlled rectifier* (s.c.r.) being one particular type. An s.c.r. with its equivalent circuit and symbol is shown in Fig. 18. With G open-circuit

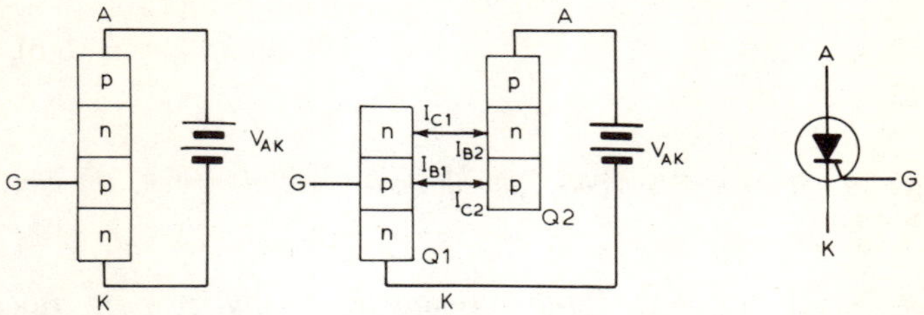

Fig. 18.—*Silicon controlled rectifier, with equivalent circuit and symbol.*

Q_1 is cut off, so the device will not conduct unless V_{AK} is made sufficiently positive for it to avalanche. If the gate is made positive I_{C1} flows. Now I_{C1} is base current for Q_2 and I_{C2} is base current for Q_1. Therefore I_{C1} brings on Q_2 and the cumulative action makes both transistors saturate, i.e. the voltage across the device falls to a minimum and it is in the condition of a closed switch. Once the s.c.r. is on, the gate voltage is not required to sustain it. In fact the gate cannot be used to turn it off, and this must be done by reducing V_{AK} to a very small voltage.

Other Types of Thyristor

Symbols for other members of the thyristor family are shown in Fig. 19. The *silicon controlled switch* (s.c.s.) differs in having a second gate, which enables the device to be switched off by gating, as well as on. To switch a thyristor on without any gate voltage the device must be made to avalanche. Such a device, which will avalanche in either direction in a controlled manner, is the *bi-directional diode* thyristor or *diac*. A gated

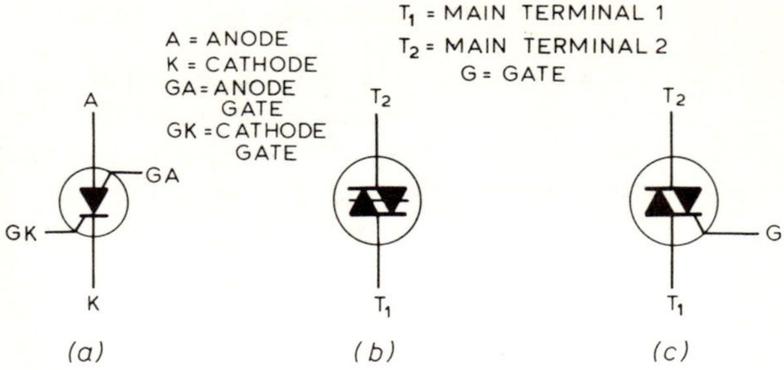

Fig. 19.—*Thyristor symbols:* (*a*) *silicon controlled switch;* (*b*) *diac;* (*c*) *triac.*

version of this is the *bi-directional triode* thyristor or *triac*, which will conduct in either direction when gated with a pulse of the appropriate polarity. In this way it performs as a pair of thyristors connected in inverse parallel.

A further device consists of a triac with a built-in trigger diac—offering triggering protection by noise, etc.—sometimes called a *quadrac*.

VARACTOR DIODE

As the reverse bias across a pn diode is increased, current carriers are drawn away from the junction; this increases the depletion layer width and reduces the junction capacitance C_j.

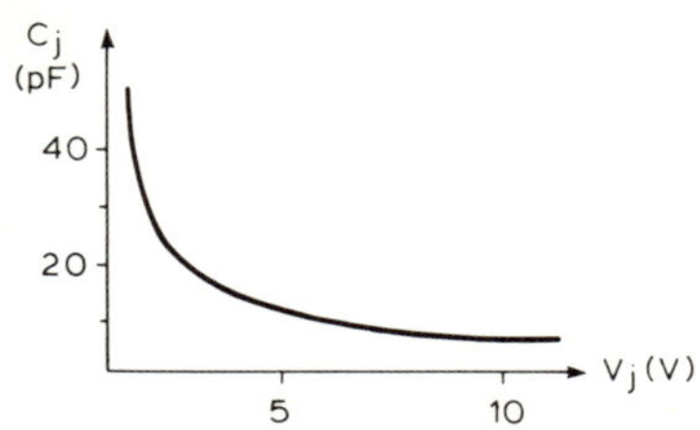

Fig. 20.—*Varactor diode characteristic.*

The *varactor diode* is a pn diode especially designed so that this voltage-dependent capacitance can be used for such applications as oscillator frequency control. A typical curve of C_j against V_j is shown in Fig. 20. Although the junction capacitance is high when the device is forward biased, it is unusual to use a varactor diode in this mode because of its loading effect on the circuit in which it is used.

This diode is variously called the varactor, varicap or variable capacitance diode.

HALL EFFECT

Applying the usual rules it can be seen that the current carriers which make up I in the semiconductor block of Fig. 21 will

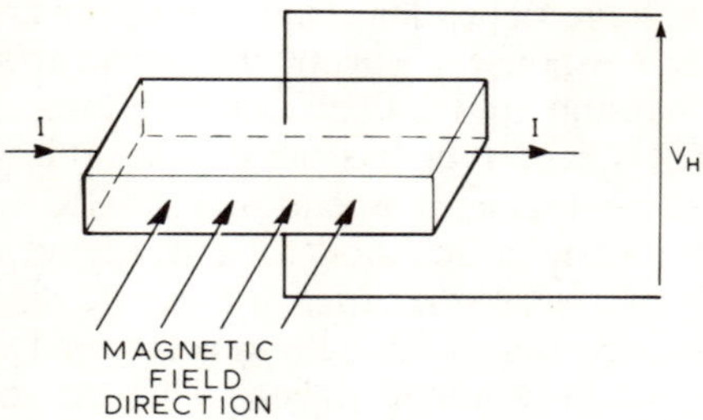

Fig. 21—*Hall effect: carriers are moved upwards, making* V_H *positive for p-type material or negative for n-type.*

be forced upwards. The uneven distribution of charge between the top and bottom surfaces of the block will give rise to a voltage V_H, a phenomenon known as the *Hall effect*. If the material is p-type the carriers will be holes and V_H will be positive, while for an n-type material V_H will be negative. The effect may be used to classify an unknown semiconductor p or n-type, to measure the doping concentration of a semiconductor block, or, probe-shaped, to investigate the strength of a magnetic field.

Suppose that the current I is produced by a signal A and the magnetic field by a signal B. Then V_H will be proportional to $A \times B$. If one signal is a.c. and the other d.c. an alternating voltage proportional to the d.c. will be produced, i.e. d.c. to a.c. conversion is achieved.

The Hall effect e.m.f. produced is usually specified in millivolts, for a particular magnetic field in kilogauss, at a particular current in milliamps.

SEMICONDUCTOR DIODES

The desirable features of semiconductor diodes, compared to thermionic diodes, are their small size and weight, absence of heater supplies, lower shunt-capacitance and lower forward resistance. The major disadvantages are their lower reverse breakdown voltage, dependence upon temperature and slower current-change response to abrupt changes in applied voltage.

Both germanium and silicon semiconductor diodes are available. The silicon type has the important advantage that it will withstand higher temperatures—200°C as compared with 75°C for germanium diodes. The silicon diode has a slightly higher forward resistance but, on the other hand, has a higher reverse resistance than the germanium type—in some cases it may be very much higher. This, combined with a similar temperature coefficient, gives a high degree of thermal stability—a factor of great importance in d.c. amplifiers. In addition, silicon diodes have a higher reverse breakdown voltage than germanium diodes. The germanium point-contact diode is also still widely used, particularly at v.h.f. and u.h.f. where specialised manufacturing techniques must be employed.

LIGHT EMITTING DIODES

The light emitting diode (LED) consists of a single piece of n-type crystal which has a metal, normally zinc, diffused into it. The lamp that is formed is based on an effect demonstrated in

1932 by Losser, who used gallium arsenide (GaAs) as the n-type crystal. When forward bias is applied to the lamp (typically at about 1·7 V) radiation is emitted due to the recombination of holes and electrons giving up excess energy in radiated form.

The nature of the radiation depends on the material of the crystal, and while gallium arsenide LEDs are still the cheapest to produce—giving infra-red or visible red light—other crystal combinations are available to give white, orange, yellow and green.

This type of lamp finds applications in systems normally using filament lamps and photodetectors; for example punch card and tape recorders, shaft encoders, alarms and control systems. The lamp can also be modulated to form the basis of short-distance communications links.

The diodes can also be combined in multiple-pin i.c.-type encapsulations to form "arrays" which, with the bias switched to the correct combination of pins, will give numeric or alpha-numeric displays.

GUNN-EFFECT DEVICES

Gunn-effect devices (in particular Gunn diodes) were discovered by J. B. Gunn in 1963. The fundamental feature of the device is that under high electric stress thin slices of gallium arsenide and indium phosphide can be made to pass a periodically-fluctuating current, i.e. they oscillate.

The frequency of oscillation can be made to lie in the r.f. microwave region when the applied voltage passes a certain critical value and for particular thicknesses of specimen. The process is based on theories of travelling domains created by negative resistances, shown to be possible by Ridley and Watkins in 1961. Briefly drifting electrons acquire energy from an applied electric field and increase their effective mass. This reduces their drift velocity towards the anode. Additionally, areas of negative charge form initially towards the cathode end of the semiconductor and the electrons travel in "drifts" through the material. When these reach the anode the domain is extinguished and the process repeats itself. The repetition rate of the domain formation and extinguishment forms the oscillation frequency.

The mode of propagation along the length of the specimen is called the "transit-time" mode and was the one first noted by Gunn. Typical operating voltages are around 6 V.

Other modes of Gunn operation have been discovered; principal amongst them is the limited space-charge accumulation (l.s.a.) mode. This uses a large steady bias and does not permit the formation of domains; the device is coupled into any resonant cavity and is used simply as the negative-resistance component required by any oscillator. With less limitation of device size (the cavity providing the resonant frequency choice) more powerful diodes can be built.

In the transit-time mode a conventional oscillator can be tuned across 0·5 per cent (say 50 MHz at 10 GHz) using a varactor diode, and a 10 per cent tuning range (say 1 GHz at 10 GHz) can be readily achieved by mechanical tuning. Efficiencies are rather higher than counterpart thermionic devices: 6 kW pulse powers are now available from l.s.a. devices at 1·5 GHz (cf. 300 W from transit-time mode) and about 3 W pulse power at 50 GHz.

4

TRANSISTOR CIRCUIT TECHNIQUES

Temperature Effects

If a d.c. voltage is applied between collector and base of a pnp transistor (see Fig. 1), with the emitter connection left open-circuit, a current can be seen to flow of the order of a few micro-amps. The collector/base junction is reverse biased so no current should flow, but this neglects the thermally produced hole/electron pairs in the basic material. As the temperature increases so does this current, and for germanium an 8 or 9°C change in temperature doubles the current. This current is notated I_{CBO}, i.e. collector/base current with emitter open-circuited.

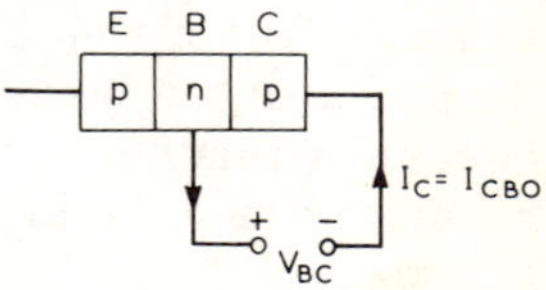

Fig. 1.—*Collector-base leakage current with emitter open-circuit.*

A second voltage source can now be considered connected between emitter and base. The current that is set up is in opposition to the leakage current in the base (Fig. 2(a)) and, at some value of V_{EB}, I_B will equal I_{CBO}.

At that voltage, the base connection is effectively performing no function and it could be removed without changing the collector current (Fig. 2(b)). The effective leakage current in this condition is termed I_{CEO}, i.e. collector-emitter current with base open-circuit. With a value of, say, 5 μA for I_{CBO}, I_{CEO} may be equal to 250 μA; the transistor is therefore producing current magnification of the leakage even without a base connection.

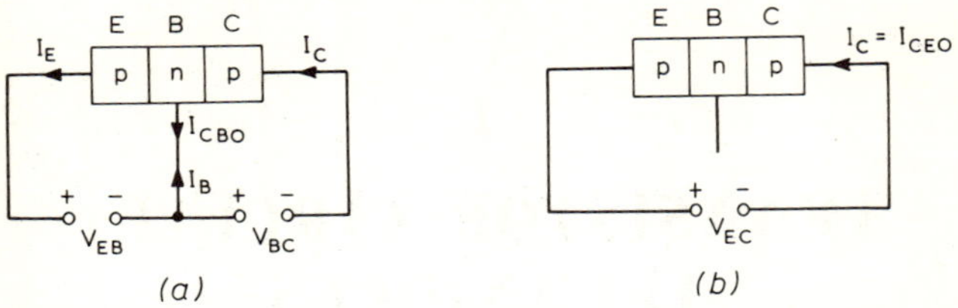

Fig. 2.—*(a) Balancing* I_{CBO} *with a base-emitter bias voltage. (b) Effective circuit when* $I_{CBO} = I_B$.

The collector current has two components, one due to the input current and one due to the leakage that is present because of thermal generation. As junction temperature increases the total collector current increases; this change in the operating point can give rise to distortion of a signal being amplified. Under extreme conditions the rise in collector current increases the junction temperature still further, which increases collector current still more.

This cyclic action can of course lead to the destruction of the transistor itself, and the process is often known as *thermal runaway.* The effect is more likely with germanium transistors than with silicon because the thermally generated current is larger in germanium.

Some form of stabilisation is needed to prevent such effects.

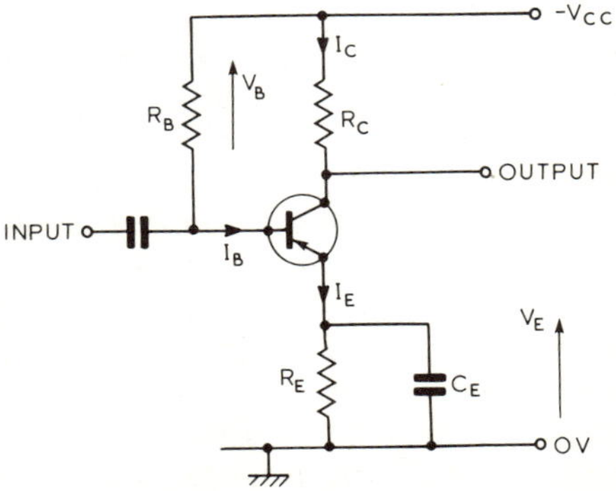

Fig. 3.—*Simple form of d.c. stabilisation for thermal effects.*

A simple form of such an arrangement is shown in Fig. 3. The effective components of the circuit are R_B and R_E. C_E effectively short-circuits R_E at signal frequencies and prevents negative feedback due to signal voltage drop across R_E.

If I_C increases due to temperature effects, I_E also increases and V_E, the voltage drop across R_E, also increases. The available bias voltage, V_B, must decrease because it equals $V_{CC} - V_E$ (neglecting the small base–emitter junction potential). The reduction in V_B reduces the base current and therefore the collector current. Any rise in collector current should therefore always be self-restricting.

The effectiveness of the circuit is low, however, because there is insufficient base-current control; it is usual to modify it to the arrangement shown in Fig. 4, where a potential

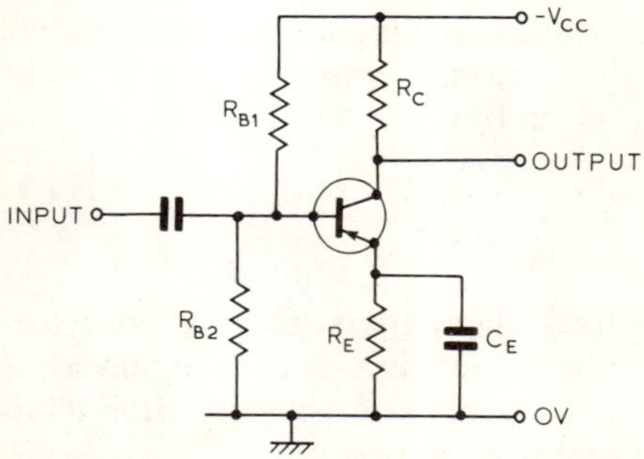

Fig. 4.—*Improved d.c. stabilisation using a potential divider arrangement on the base.*

divider holds the base to a more constant potential, the divider chain taking a current several orders greater than the base current.

This implies small values for R_{B1} and R_{B2}. But small values of these resistors present disadvantages:

(1) The input signal is effectively shunted by the combined resistance of R_{B1} and R_{B2} so reducing gain.

(2) The d.c. current drain through R_{B1} and R_{B2} may be excessive.

Absolute Maximum Ratings

In general manufacturers of transistors give ratings for their devices that must never be exceeded; in design the spread of transistor supply voltage and other circuit parameters must always be taken into account to ensure that none of the ratings is exceeded.

It is important to note that it must not be assumed that operation below the limit of one rating parameter does not necessarily allow the operation above another limit.

Dissipation

The currents flowing in a transistor—principally the collector current—produce heating effects in the regions of differing resistance. The junction temperatures of the transistor must not rise above specified limits or else the transistor will fail.

The equilibrium that junction temperature will reach will depend on the ambient temperature, and total permissible dissipation is given by:

$$P_{tot} = \frac{T_{j\max} - T_{amb}}{\theta}$$

where P_{tot} = total dissipation (usually collector dissipation unless there is a large base power)

$T_{j\max}$ = maximum permissible junction temperature

T_{amb} = ambient temperature

θ = "thermal resistance", i.e. rise in junction temperature above ambient for every unit of power dissipation.

For power transistors, heat sinks are normally provided to aid dissipation by conducting heat away from the transistor. The ability of a heat sink to remove heat is defined in terms of its thermal resistance θ_H. However the transistor itself has thermal resistance in its ability to transfer heat between junction and its mounting case (θ_m) and there may be an insulating washer between transistor and heat sink which also has thermal resistance (θ_i).

The total thermal resistance of a power transistor mounted on a heat sink with an insulating washer would therefore be $\theta_m + \theta_i + \theta_H$, and the total dissipation would be given by:

$$P_{tot} = \frac{T_{j\max} - T_{amb}}{\theta_m + \theta_i + \theta_H}$$

θ_m is of course fixed by the transistor manufacturer, who keeps the figure as low as possible; the other parameters can be modified by the circuit designer: θ_H may be reduced by using a larger heat sink, for example, or one with forced air cooling, or one painted matt black, while θ_i may be reduced by using thinner washers and by ensuring good thermal contact between washer and transistor, and washer and heat sink, using a silicon jointing compound.

As an example, a transistor of $\theta_m = 3{\cdot}0$°C/W using a washer of $\theta_i = 0{\cdot}5$°C/W and a heat sink of 4·5°C/W might have a maximum junction temperature of 110°C. If the collector dissipation was 10 W,

$$10 = \frac{110 - T_{amb}}{3{\cdot}0 + 0{\cdot}5 + 4{\cdot}5}$$

so the permissible ambient temperature under these conditions would be

$$T_{amb} = 110 - 10 \times 8{\cdot}0 = 30\text{°C}$$

High Frequency Limits

To define the high-frequency performance of a transistor, two parameters are available. One is known as the transition frequency (f_t) and the other as the alpha cut-off frequency (f_{ab}).

The transition frequency, f_t, is defined as the extrapolated unity gain frequency—the frequency at which the gain of a circuit in common-emitter is unity. It is often extrapolated because of the very high frequencies involved in modern transistors. A frequency quoted at unity gain is also the same thing as the gain–bandwidth product.

The alpha cut-off frequency, f_{ab}, is also sometimes called the common-base cut-off frequency (f_{hfb}); it is the frequency at which the current gain in common-base falls by half.

An idea of the maximum frequency of operation of a transistor is essential if it is required to achieve gain from that device. It would not, for example, be possible to operate a transistor whose f_t was 250 MHz at a frequency of 50 MHz and expect to get a gain greater than 5.

Simplified A.C. Circuits

Common-emitter, common-base and common-collector circuits are shown in Figs. 5, 6 and 7 respectively. In Fig. 5:

$$R_{in} = h_{ie} \approx 1\ \text{k}\Omega$$
$$R_{out} = R_c$$
$$\text{Current gain} = h_{fe}$$
$$\text{Voltage gain} = g_m R_c$$

In Fig. 6:

$$R_{in} = h_{ie}/h_{fe} = 1/g_m$$
$$R_{out} = R_c$$
$$\text{Current gain} = 1$$
$$\text{Voltage gain} = g_m R_c$$

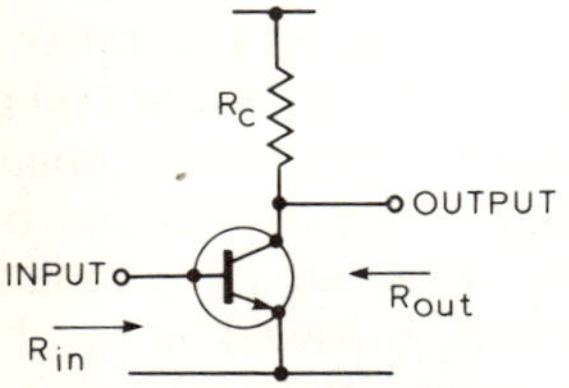

Fig. 5.—*Common-emitter circuit.*

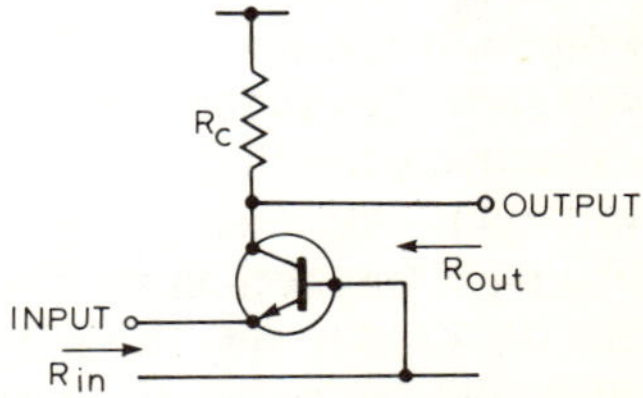

Fig. 6.—*Common-base circuit.*

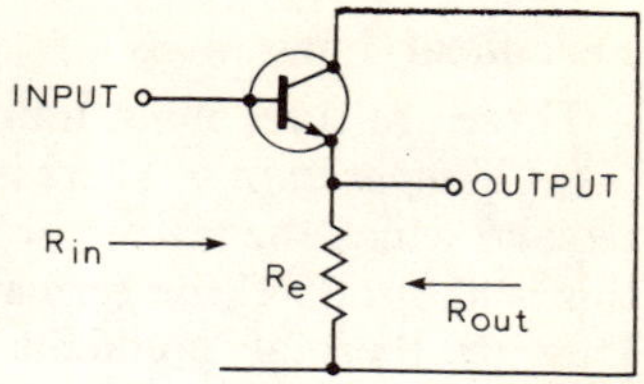

Fig. 7.—*Common-collector (emitter follower) circuit.*

In Fig. 7:

$$R_{in} = h_{fe}R_e + h_{ie}$$

$$R_{out} \approx 1/g_m \text{ when source resistance} = 0$$

$$R_{out} = R_s/h_{fe} \text{ when source resistance} \neq 0$$

$$\text{Current gain} = h_{fe}$$

$$\text{Voltage gain} = 1$$

Note that in each case $g_m = h_{fe}/h_{ie}$.

GENERAL

The semiconductor materials used in current transistors may be either germanium or silicon, though in the future other materials such as gallium arsenide and gallium arsenide phosphide will be of importance. For industrial applications, silicon has the advantage over germanium of working up to higher temperatures. Both pnp and npn transistors are in common use, and must be carefully distinguished: whereas the pnp transistor has a negative potential applied to its collector, and positive current bias is required, the collector potential of an npn type is positive and it requires negative current bias. In circuit diagrams, the arrow head on the emitter electrodes of a pnp transistor is usually shown pointing *towards* the base symbol; on an npn type it points *away* from the base symbol. In general it is easier to construct an npn silicon transistor and a pnp germanium transistor.

Functional Limitations

There are three main limitations on transistor operation:

(a) *Temperature.*—There is a maximum junction temperature beyond which the transistor junctions break down irreversibly; this is about 85°C for germanium and about 200°C for silicon. Due to thermal gradients in the structure, however, the permissible ambient operating temperatures are lower than this. Two undesirable effects of temperature rise are increase of effective emitter conductance and of the reverse leakage current. Silicon transistors are less affected by heat because the energy gap—the energy that an electron requires in order to break free from its parent atom—is greater with silicon than germanium (1·08 eV for silicon, 0·72 eV for germanium).

(b) *Frequency.*—The gain available from a transistor decreases as the frequency increases. This limitation is due to: (1) Migration time for charge carriers from the emitter crossing the base to the collector. This results in a decrease of the current gain α. (2) The low-pass filters formed by the resistance of the material combined with a number of capacitances, associated with the diffusion process of charge carriers through the base and with the junction potential barriers. The upper-frequency cut-off value of a transistor is usually quoted as the transition frequency (f_t) which is the frequency at which the common-emitter current gain (β) falls to unity.

(c) *Voltage and Current.*—To reduce the migration time of charge carriers in the base, the base region is made physically very thin. This means that breakdown at the junction occurs at relatively low voltages and so transistor operating voltages tend to be low. At any given operating voltage the current is limited by the permissible power dissipation at the collector junction in order that the maximum operating temperature is not exceeded.

5

AMPLIFIERS*

It is convenient for purposes of classification to divide amplifiers into voltage and power amplification stages: in a multi-stage amplifier, the initial stages are concerned with voltage amplification, i.e. the amplification of a small signal voltage, and the last stage with supplying power to a load of some kind.

VOLTAGE AMPLIFICATION

The most widely known form of voltage amplifier is the Class A amplifier, defined as one in which the grid/base bias and the a.c. signal voltage applied are such that anode/collector current flows at all times. The operation of a triode valve and a transistor as simple amplifiers has already been described in Sections 2 and 3. The output of a stage of amplification may be passed on to the next stage by means of a transformer or a resistance-capacitance network. Because of expense and the difficulties of transformer design for low frequency signals, it is usual to use the latter method. A two-stage resistance-capacitance coupled voltage amplifier using two triodes is shown in Fig. 1(a) and using two transistors in Fig. 1(b). The output of the first stage is developed across its load resistor and passed on via the capacitor to the second stage. The capacitor permits alternating potential to pass through it but blocks the way to direct current. Thus it prevents the supply voltage reaching the grid/base of the second device. The gain of each valve stage is given by the formula

$$G = \frac{\mu R_L}{R_L + r_a} \qquad \text{(1a)}$$

* Unless otherwise indicated in the text, this section refers to both valve and transistor circuits.

where μ is the amplification factor of the valve, R_L is the anode load in ohms and r_a the anode impedance of the valve.

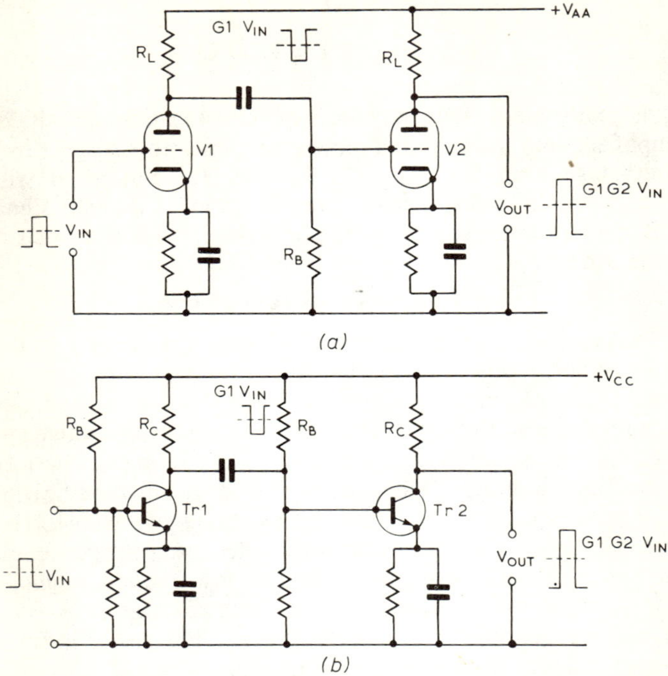

Fig. 1.—*Resistance–capacitance coupled two-stage voltage amplifier: (a) using triode valves; (b) using transistors.*

The gain of each transistor stage is given by the formula

$$G = \frac{h_{fe}}{h_{ie}} . R_C \tag{1b}$$

where h_{fe} is the small signal current gain of the transistor in common emitter, h_{ie} is the common-emitter input impedance in ohms, and R_C is the collector load.

The total gain of the two stages will be

$$G_t = G_1 . G_2 \tag{2}$$

where G_1 and G_2 are the gains of each stage calculated from equations (1a and b).

Many engineers talk about the stage gain of a voltage amplifier in decibels although this is not strictly correct since if the input and output impedances are not equal voltage gain or loss should be specified with respect to the impedance.

If the gain of one stage, G_1, is calculated from equation (1a and b) the gain of that stage in decibels is given by

$$\text{Gain in dB} = 20 \log_{10} G_1 \qquad (3)$$

In the same way the gain of the second stage can also be calculated in dB, and since logarithms can be added directly to one another, the total gain of the two stages is the sum of their individual gains. If a third stage is added to the amplifier then the gain would increase by a further number of decibels equal to the gain of the new stage.

Another feature, of great importance in electronics, is the change of phase that occurs in a stage of grounded-cathode valve or common-emitter transistor amplification. In one stage the signal is changed from, say, a positive going signal at the grid or base to a negative going signal at the output. Therefore if an amplified signal of the same polarity is required it is necessary to use two stages of amplification, even though one stage alone might have provided the required gain.

Valve Biasing

Negative bias is applied to an amplifying valve to make it operate on the linear portion of its grid voltage/anode current characteristic curve. It may be applied between the grid and cathode by causing the cathode to be positive with respect to the grid. The usual method of doing this is by inserting a resistor in the cathode lead. When current is flowing through the valve, a voltage will be built up across this resistor, making the cathode positive with respect to earth. By returning the grid circuit to earth, the grid is made negative with respect to the cathode. A capacitor across the resistor provides a low impedance to earth for the a.c. component of the anode current.

If the value of the resistor is carefully chosen, any value of bias voltage can be obtained, the polarity being as shown in Fig. 7, Section 2. The actual signal is unaffected by this bias network provided that the bypass capacitor is of such value that it offers negligible reactance to the signal.

The value (in ohms) of a cathode bias resistor may be calculated from the formula

$$R\text{ (cathode)} = \frac{\text{grid bias voltage required (Volts)}}{\text{total cathode current (mA)}} \times 1{,}000$$

and the value of the associated bypass is given approximately by

$$C \simeq \frac{1}{6{\cdot}28 \times f \times (R/10)}$$

where C is the bypass capacitor in farads, f the lowest frequency to be bypassed and R the resistance across C. It should be noted that the total cathode current is the sum of currents flowing in all grid and anode circuits.

Inter-stage Coupling

As has already been mentioned, it is usual to provide inter-stage coupling between two stages of voltage amplification by means of a resistance-capacitance network as shown in Fig. 1(a). The signal applied across the grid of the second valve is developed across the resistor R_B. The capacitor in series with this resistor isolates the h.t. voltage at V1 anode from V2 grid. In audio amplifiers this capacitor is usually about 0·5 μF, but in general this capacitor must be given a value according to the job the amplifier is to do. If pulses are being amplified the value must be chosen with care so that the time constant of the resistor and capacitor is not too great and the amplifier "blocked off" when the second pulse arrives. Again, it is sometimes necessary to amplify signals which change at a very low frequency, or pulses which are very long in time. In these cases, to avoid distortion a large value of capacitance is required. To find a large-value capacitor which will withstand an h.t. voltage and have a high resistance to leakage is sometimes difficult, as capacitors satisfying this requirement are expensive. The solution when amplifiers are designed for very low frequency work often lies in using d.c. amplifier techniques.

Transistor Amplifiers

There are important differences between valve and transistor amplifiers. In a thermionic valve the anode current is controlled by variation of the potential of the control grid. Although in certain circumstances the control grid may draw current, this does not contribute to the fundamental action of the valve and, in most cases, merely has a nuisance value. In the case of the transistor, however, control of the collector current is brought about by variations of the current in the emitter-base path, and this current flow is fundamental to the operation of the transistor. This is the most significant difference between valves and transistors so far as circuitry is concerned: the need for current in the control circuit of transistors must always be recognised when devising amplifiers or

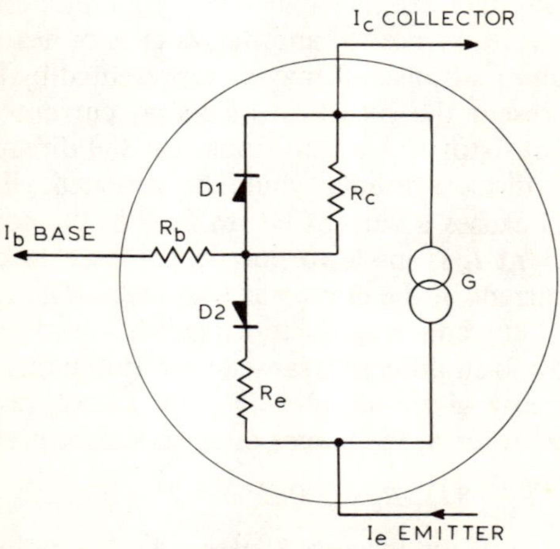

Fig. 2.—*Representation of the action of a transistor.*

oscillators around them. A further difference between valves and transistors is that the input impedance of a transistor amplifier of the kind shown in Fig. 3 is much lower than that of the equivalent valve amplifier. It is only of the order of a hundred ohms as opposed to many thousands of ohms in the case of the valve input. This limits its use where a circuit

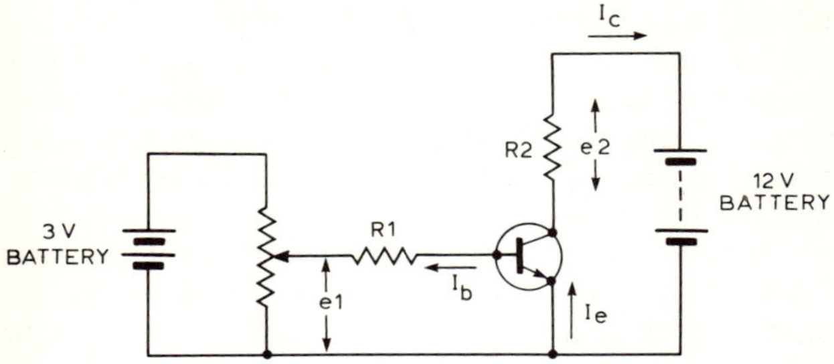

Fig. 3.—*Common-emitter transistor amplifier.*

must not load a previous stage. The great advantages of the transistor amplifier are its small size, long life and robustness, the low d.c. voltages needed and the absence of heater supplies.

The action of a transistor may be represented by Fig. 2. The arrows represent the direction of actual current flow in an npn-type transistor: for a pnp transistor the directions of the arrows and diode symbols would be reversed. The current generator G causes a current βI_b to flow in the collector lead when a current I_b is made to flow in the base lead. Accordingly, the current in the emitter is $(\beta+1)\, I_b$. The number β is called the "current amplification factor" and varies from about 20 to 450 in different transistors but remains reasonably constant in any given sample. D1 and D2 represent perfect diodes. The order of the values of the resistances are:

$$R_e \sim 5\ \Omega,\ R_b \sim 500\ \Omega \text{ and } R_c \sim 500\ k\Omega$$

The current in R_c is discussed later. At this point it can be assumed that this current is negligible compared with I_b.

Consider the circuit of Fig. 3. The base current

$$I_b = \frac{e_1}{R_1 + R_b + R_e} \simeq \frac{e_1}{R_1 + R_b}$$

since R_e is very small compared with R_b. A typical value of I_b in a small transistor is 0·5 mA; therefore if $e_1 = 1{\cdot}5$ V then $R_1 + R_b = 1{\cdot}5$ V/0·5 mA = 3 kΩ.

$$\therefore R_1 = 2{,}500\ \Omega$$

If the current gain of this transistor is 20 then $I_c = 10$ mA. Let $R_2 = 680\ \Omega$, making $e_2 = I_c . R_2 = 6{\cdot}8$ V.

In this circuit,

$$\text{Current gain} = I_c/I_b = 20.$$
$$\text{Voltage gain} = e_2/e_1 = 6{\cdot}8/1{\cdot}5 = 4{\cdot}5$$
$$\text{Power gain} = 20 \times 4{\cdot}5 = 90$$

If R1 and R2 are replaced by suitable input and output devices, and e_1 is taken from a tapping on the 12 V battery then, in principle, one transistor forms a complete amplifier circuit having a useful gain. As no heater is present the amplifier functions immediately the power is switched on.

Spread in Characteristics and Thermal Effects

The arrangement shown in Fig. 3 is not adequate for a practical amplifier. The value β can vary from 150 to 450 in transistors from the same production run. The manufacturer measures β (small signals) and divides the transistors into groups, e.g. β from 150 to 250, 250 to 350, and 350 to 450. These groups are then given different type numbers. The equipment designer must accept the spread of β in a given type and must arrange his circuit so that it will operate with any transistor from a given group. As an illustration of this factor, suppose that the β of the transistor in Fig. 3 is 50 instead of 20 and that the other details of the circuit are unchanged. Calculating as before gives $e_2 = 17$ V, which is impossible as the supply voltage is only 12 V.

In practice e_2 becomes 12 V, the collector-to-emitter voltage becomes a few millivolts and the output circuit becomes completely insensitive to any change in the input current until I_b is reduced to the point where e_2 becomes less than 12 V, i.e. I_b is less than 0·35 mA. It is possible to overcome this difficulty by placing a variable resistor in series with R1, and trimming it to suit the transistor.

The spread in β, however, is not the only difficulty encountered with transistors. The most serious disadvantage is that the quantities R_e, R_c and R_b vary considerably with temperature, in particular the value of R_c decreases rapidly with increase in temperature. Consider Fig. 2 assuming that the collector is connected to -12 V, the emitter to zero volts and the base is left open circuit. The current flows from the emitter lead through R_e, D2, R_c to the collector lead. $R_c \gg R_e$ so that the current is $12\ \text{V}/R_c = 24\ \mu\text{A}$ when $R_c = 500\ \text{k}\Omega$.

The current of 24 μA, flowing through D2, is seen as a signal current so that G will generate $24\ \mu\text{A} \times \beta$. If $\beta = 50$, the total collector current is $50 \times 24\ \mu\text{A} = 1{\cdot}2$ mA. The heat developed in the transistor is $12\ \text{V} \times 1{\cdot}2\ \text{mA} = 14{\cdot}4$ mW, which will raise the temperature and decrease R_c and thus further increase the dissipation. The process is cumulative and may result in the transistor burning out. The effect is known as "thermal runaway" and precautions must be taken to prevent this from occurring. Some resistance in the collector circuit will impose a limit on dissipation in the transistor and prevent a runaway. The technique usually adopted for overcoming the "thermal runaway" effect is described in Section 4. In any circuit the changes in R_c, R_b and R_e give rise to unwanted d.c. signals and steps must be taken to minimise their effects. Changes in transistor parameters come under the general heading of "thermal drift" and present a big problem in circuit design using germanium transistors and in high-temperature usage of silicon transistors.

Transistor Parameters

When transistors are used as amplifiers, constant voltage biasing is used to ensure that the transistor works on a desirable portion of its characteristics. Operation as an amplifier therefore depends upon the a.c. impedances presented to the electrodes by the circuit elements and on the slope of the tangent to the characteristic curves at the operating points. There is a wide variety of curves and slopes which can be specified for any one transistor, but only four independent values are necessary to completely define its operation. In the interests of standardisation there is a general agreement to specify a set of four values which are known as the "*h*" parameters; the four are of mixed dimensions, and "*h*" was chosen as a symbol to represent "hybrid". These four values are listed below:

$h_{11}=\frac{v_1}{i_1}$ Variation of input current with input voltage. This is the input impedance with output short-circuited, and its value is in ohms.

$h_{21}=\frac{i_2}{i_1}$ Variation of output current with input current. This is the current gain with output short-circuited, and is a pure number.

$h_{22}=\frac{i_2}{v_2}$ Variation of output voltage with output current This is the output admittance with input open-circuited, and its value is in mhos.

$h_{12}=\frac{v_1}{v_2}$ Variation of output voltage with input voltage. This is the voltage feedback ratio with input open-circuited, and is a pure number.

Each *h* parameter will change according to the mode of transistor operation. For most purposes only h_{11} and h_{21} are required, and for common-emitter operation these are designated h_{ie} and h_{fe} respectively. Likewise the parameters for common base are known as h_{ib} and h_{fb}.

Remember, from Section 4, that h_{FE} is the static forward-current transfer ratio in common-emitter, and is otherwise known as the d.c. β or d.c. current gain. h_{fe} is the small-signal short-circuit forward-current transfer ratio in common emitter (a.c. β).

DISTORTION

If the amplification factor or gain of an amplifier varies with frequency the result is "frequency distortion". It can usually be corrected: for example, an audio amplifier whose gain may begin to decrease at say 300 kHz can be made to give satisfactory gain at 3 MHz with compensating chokes in the load.

Another form of distortion common to all amplifiers is non-linearity. If the input increases in steps, the output should increase in equal steps multiplied by the gain. If this is not the case, the amplification is said to be non-linear. Many amplifiers have this problem but good design can largely overcome it. Valves should not be overdriven, and should be chosen with care. In transistor amplifiers, variations in input impedance can produce a non-linearity that is harder to solve, although variations in input impedance can sometimes be masked by deliberately producing an inter-stage mismatch.

BANDWIDTH

Amplifiers are designed to work over a given band of frequencies beyond the limits of which the gain falls off. The bandwidth of an amplifier is commonly quoted as that between the half-power (−3 dB) points.

Amplifiers containing resonant circuits tuned to the same frequency or frequency selective filters normally have a narrow bandwidth of a few per cent of the centre frequency. Such amplifiers are termed *narrowband* or *selective* amplifiers. Highly selective amplifiers are required in applications where it is necessary to distinguish between the signal being amplified and background noise. Amplifiers capable of providing consistent gain over an appreciable range of frequencies are termed *wideband* amplifiers.

A signal of basically sine-wave characteristics can be amplified satisfactorily in a narrowband device. Low frequency rectangular pulses and square waves, on the other hand, can be shown to contain odd harmonics beyond the twenty-fifth and such signals would be severely distorted by a narrowband amplifier.

POWER AMPLIFIERS

The main difference between the power amplifier and the voltage amplifier lies in the method of coupling the load to the anode or collector. To transfer maximum power from any generator to a load it is necessary that the generator impedance and the load impedance should be equal. Since an amplifier can be regarded as a generator with an internal impedance that can readily be calculated, it is possible to satisfy this condition simply by including between the load and the output a matching transformer. A common example of this is the output stage of a valved radio receiver where the last stage is matched by a transformer to the loudspeaker.

In Fig. 4 a pentode valve is shown connected to a load via a transformer. This type of amplifier is usually a compromise because if the load were correctly matched, as has been suggested, the effect on the valve's dynamic behaviour would be such as to introduce distortion. Therefore it is usual to deliberately mismatch the system whilst aiming at getting the most power out with the least distortion.

The transistor has proved itself very useful as a power amplifier when care is taken to see that it is not over-rated and asked to dissipate more heat than it is capable of doing—a common cause of failure in transistors.

The basic circuit for the transistor power amplifier does not differ greatly from the one for a voltage amplifier, except that the components are rated for higher currents. A better match can sometimes be achieved with transistors than with thermionic valves and a higher efficiency obtained. A popular arrangement for an audio output stage is shown in Fig. 5. The transistor amplifies the input current and drives βI_{in}

through the load, developing a voltage $\beta I_{in} Z$ across it; and in this respect is similar in action to the pentode valve. Any increase of collector and emitter current caused by the tran-

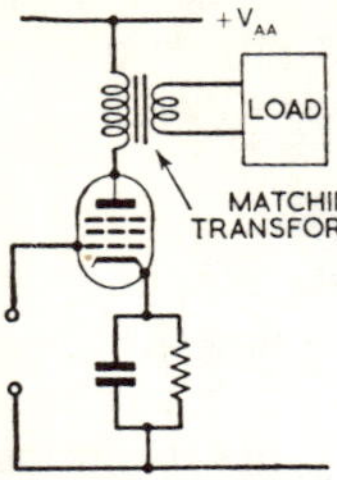

Fig. 4.—*Pentode valve used as a power amplifier.*

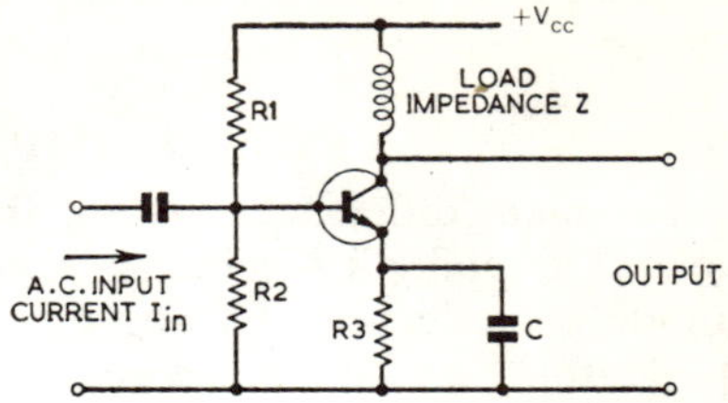

Fig. 5.—*Basic transistor amplifier output stage.*

sistor warming up produces an increased voltage across R3 which in turn causes the emitter d.c. potential to approach the potential established by the resistor chain R1 and R2. This has the effect of reducing the d.c. input signal to the base and reducing the collector and emitter currents. Thus the circuit is stabilised against thermal effects. The purpose of the capacitor C is that it shall act as a shunt across R3 and bypass any a.c. signals present so that the maximum gain may be achieved from the circuit. This gain is proportional to the β of the transistor.

Push-pull Amplification

To increase the output power of an amplifier, the output valves may be connected in parallel or push–pull. The basic push–pull arrangement is shown in Fig. 6. The advantages of push–pull operation are discussed under the heading "Class B Operation" later. In the push–pull circuit, oppositely phased signals are applied to the control grids of each valve. These signals may be provided by a centre-tapped transformer or a phase splitter stage.

At any instant when V1 grid voltage is becoming less negative V2 grid voltage is becoming more negative. The grids are

Fig. 6.—*Basic push–pull output circuit, using valves. (Note: inputs and* V1 *grid and* V2 *grid must be out of phase.) Transistors may also be used in push–pull arrangements.*

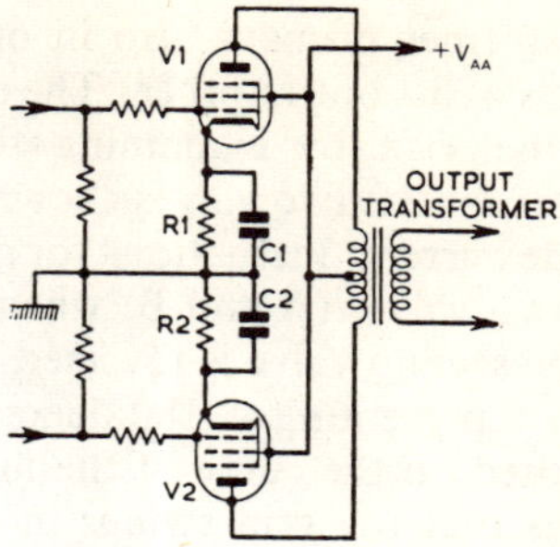

always in opposite phase so that while the anode current in V1 is increasing that in V2 is decreasing. A centre-tapped output transformer is used, its primary winding being connected between the anodes of the two valves so that the changes in

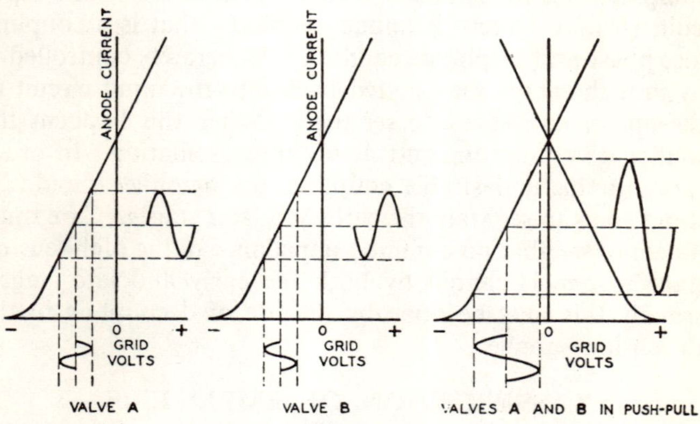

Fig. 7.—*Effect of the grid swings of two valves connected in push–pull on the resultant anode current.*

the currents flowing in the two valves are made to produce additive effects in the secondary winding of the output transformer. This result is achieved because, although the current changes at any instant in the two valves are in opposite directions, these currents pass through the two halves of the transformer primary also in opposite directions so that their effects assist one another. The effect of push–pull amplification can be seen by examining the grid voltage/anode current curves for the two valves shown in Fig. 7. (a) and (b) show the anode current fluctuations for given grid voltage fluctuations of the two valves (A and B) when considered alone. If the curve of the second valve is reversed by rotating it about the vertical axis, representing a 180-degree phase shift, and then superimposed on the curve of the first valve the result, as shown in (c), is that the grid swings in the two valves when operating simultaneously assist one another in their effects on the resultant anode current.

Stability of Amplifiers

The voltages and currents involved in the output circuits of an amplifier are much larger than those involved in the input circuit, so that if there is undue coupling—that is a coupling whose phase and amplitude cannot be deliberately controlled—then enough energy may be fed back into the input circuit to make up for any ohmic losses there. When this happens the amplifier may burst uncontrollably into oscillation. In order to prevent this undesirable coupling, the amplifier should be laid out so as to separate the early and later stages. Care must be taken to see that no common impedance in the high tension voltage system is shared by both the early and late stages. Normally this may be done by the use of decoupling in the high tension supplies.

CLASSIFICATION OF AMPLIFIERS

Valve amplifiers are classified according to the conditions governing the flow of either anode or collector current, i.e. the setting of the grid/base bias and the amplitude of the a.c. voltage applied to the control grid. Transistor amplifiers are classified in the same manner. The main classes are A, B and C, the operation of them being shown in Fig. 8. There are also intermediate classes, e.g. AB_1 and AB_2 (the suffixes indicating whether or not grid current flows in a valve amplifier).

Class A Amplifiers.—In these the bias and a.c. signal voltages are such that anode or collector current flows at all times.

Class B Amplifiers.—In these the bias is approximately equal to the cut-off value, so that the anode or collector current is approximately zero when there is no signal voltage applied,

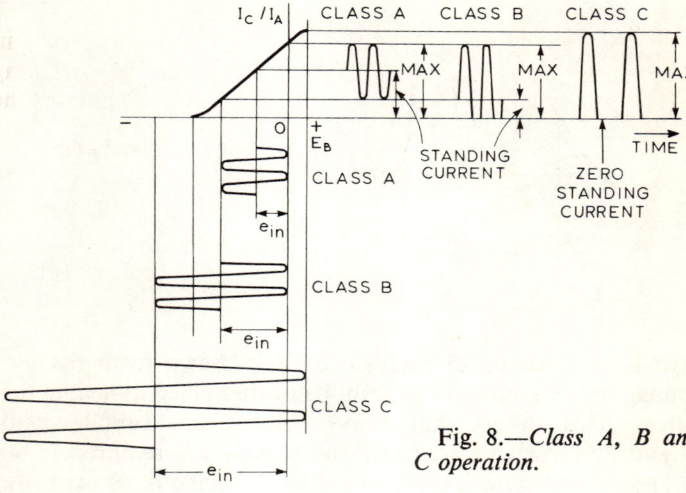

Fig. 8.—*Class A, B and C operation.*

output current flowing for approximately half of each cycle while a signal voltage is being applied.

Class C Amplifiers.—In these the bias is appreciably greater than the cut-off value. Output current therefore only flows during appreciably less than one-half of each a.c. cycle applied to the control grid or base.

Class A Operation

Most voltage amplifiers operate in Class A, and this mode is also used for power amplification where good linearity is required. With valves, the device is biased to near the mid-point of the linear portion of the I_a–V_g characteristic as shown in

Fig. 9. It should be noted that the mean anode current of a linear Class A amplifier does not change with amplitude variations of the input.

The objection to Class A operation for power amplification lies in the poor efficiency; the power delivered to the external

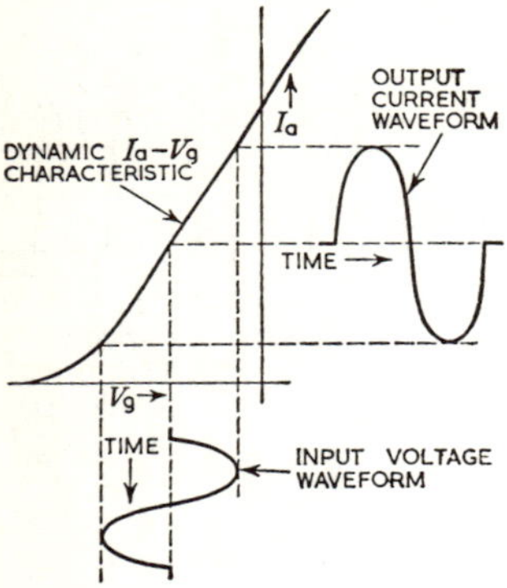

Fig. 9.—*Class A operation.*

load is only about 25 per cent of that drawn from the h.t. source, the remaining 75 per cent being dissipated as heat in the valve. This means that heavy current h.t. supplies and physically large valves or artificial cooling are required.

The theoretical efficiency of a Class A stage is 50 per cent. The reason for the great discrepancy between practical and theoretical efficiencies can be seen from Fig. 10, which shows a set of I_a–V_a characteristics for a triode and a load line A–B for a transformer-coupled load. The quiescent point is at O, the ordinate through which intersects the V_a axis at C, corresponding to the h.t. voltage. As shown, there is no very great difficulty in swinging the anode current between zero and twice the quiescent value E, but the lowest instantaneous anode potential reached is D, which represents an appreciable positive value. The anode potential cannot be swung to zero without making the grid appreciably positive with respect to the cathode, and such operation introduces considerable distortion. If the grid is kept negative with respect to the cathode throughout the cycle of the input signal, the peak value of the anode-potential swing is limited to approximately one-half of the h.t. supply, and the efficiency is only one-half of the theoretical maximum.

Somewhat higher efficiencies can be obtained by using two valves in a Class A push–pull arrangement: that is to say,

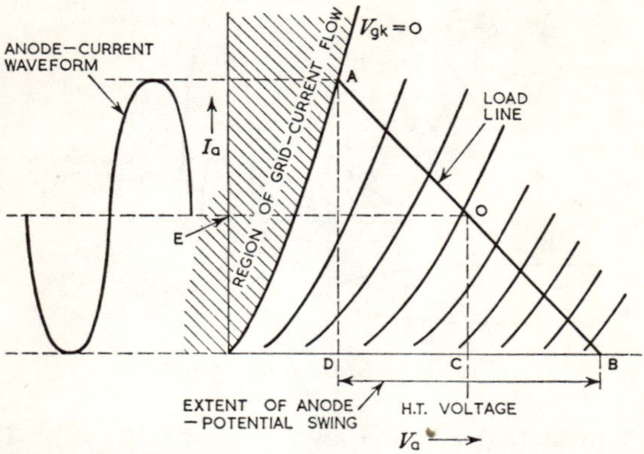

Fig. 10.—*Limitation of anode potential swing by grid current flow in a Class A amplifier.*

signals of opposite phase are applied to the grids and obtained from the anodes. However, when valves are connected in push–pull, it becomes possible to operate them in Class B or in an intermediate mode between Class A and B and to greatly increase efficiency on this account.

Class B Operation

The grid bias in a Class B amplifier is approximately equal to the cut-off bias (Fig. 11), and the input signal is great enough to cause considerable grid current. Thus the valve takes zero or very little anode current in the absence of an input signal, and the mean current rises when the signal is applied. The power taken from the h.t. supply—and the power dissipated as heat in the valve—both increase when the input signal is applied, this behaviour contrasting with that of a Class A amplifier.

The dynamic I_a–V_g characteristic of a single valve has, however, two regions of curvature which in practice cause distortion. The "bottom bend" near the point of anode-current cut-off is one such region, and the distortion caused by

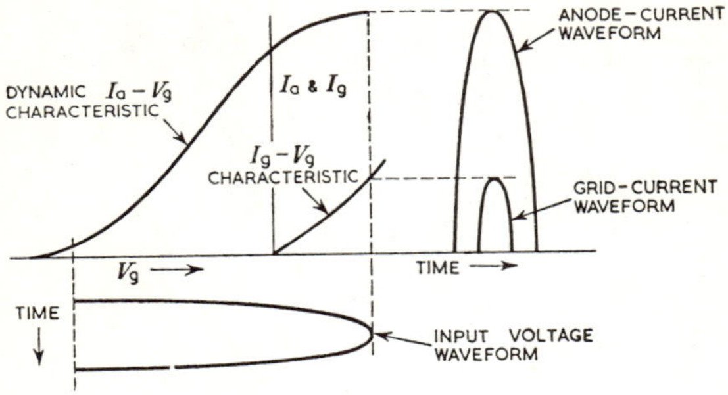

Fig. 11.—*Class B operation.*

it is so serious that Class B amplifiers seldom, if ever, use a single valve. The effects of this curvature are to a large extent eliminated by using two valves in push–pull. The push–pull connection cancels even harmonics and results in a less-distorted output waveform. This is illustrated in Fig. 12, which shows two I_a–V_g curves for the individual valves of a push–pull pair. The effective characteristic for the stage can be obtained by adding the ordinates of the two curves, and is illustrated by the dotted line. This "composite characteristic" is clearly a better approximation to the ideal than either of the individual curves.

The second region of characteristic curvature, the "top bend", occurs as the grid voltage becomes less negative, particularly with grid current. This, in flowing through the resistance of the input-signal source, generates a voltage which opposes the input signal in polarity and reduces the effective input to the valve. Thus, as the grid is driven increasingly positive, the anode current falls more and more short of the expected value and the non-linearity becomes more acute. In order to obtain high efficiency the anode potential must fall instantaneously almost to zero on positive peaks of the input signal. As shown

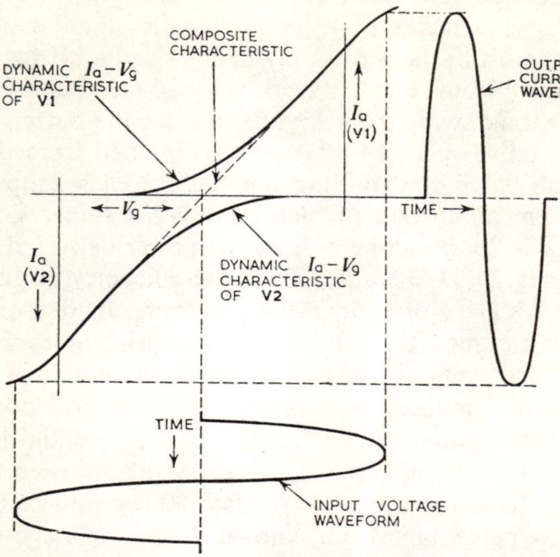

Fig. 12.—*Operation of a Class B push–pull amplifier.*

in Fig. 10, this can be achieved by driving the grid considerably positive with respect to the cathode. Thus considerable grid current and the consequent "top bend" characteristic curvature are inevitable. This source of distortion is minimised in practice by arranging for the grid-signal source to be of low resistance so that the potential difference developed by grid current, in flowing through the source, is small compared with the signal amplitude. In general, the signal source resistance is made approximately one-fifth the instantaneous input resistance of the Class B amplifier for a peak-amplitude input signal.

The fact that a Class B amplifier takes grid current implies, of course, that the previous amplifying stage must supply appreciable power. The regulation of the grid bias supply must also be considerably better to cope with the changes in input conditions.

The theoretical maximum efficiency of a Class B amplifier can be calculated in the following way. Suppose the two push–pull valves have linear I_a–V_g characteristics and are biased precisely to cut-off, the h.t. supply being V volts. If the anode currents swing from zero to a peak value of I amperes and the anode potentials swing from V volts to zero, the output power is $\frac{1}{2} \times IV$ watts. This is the total power obtained from the two valves, each valve contributing one-half of each output sine wave. The mean current of each valve is I/π amperes, for the whole stage is $2 \times I/\pi$ amperes, and the power delivered by the h.t. supply is $2 \times IV/\pi$ watts. Thus the efficiency is given by $(IV/2) \times (\pi/2 \times IV) = \pi/4$ or 78·54 per cent. In deriving this figure it is assumed that the anode potential of each valve swings to zero volts. If this occurred in practice, the anode potential would momentarily be less than the grid potential, which is at its positive peak when the anode potential is at its negative peak. The input voltage is generally adjusted to give an anode potential swing equal to, say, 80 per cent of the h.t. voltage, this percentage being known as the "voltage utilisation factor". This reduction in anode swing causes an equal reduction in efficiency, which, for a "voltage utilisation factor" of 80 per cent, is thus $0{\cdot}8 \times 78{\cdot}54 = 63$ per cent. This is the efficiency for sinusoidal input signal of constant amplitude. For such signals, however, Class C amplification can be used to give still higher efficiencies, and Class B operation is generally used for signals where the amplitude is constantly varying, depending on the nature of the input signal.

If the valve characteristics were straight, the efficiency would be independent of the input signal, the mean anode current being always a constant fraction of the peak value. Because of the "bottom bend", however, it is difficult to assess the precise value of grid potential at which the anode current is zero and it is usual to operate the valves with a small standing anode current. For small inputs, therefore, the operating conditions approximate to those of a Class A amplifier and the efficiency is low. For larger inputs, the efficiency rises to a maximum of, say, 63 per cent as calculated above. In practice, the signal amplitude varies, and the efficiency varies with it. The average efficiency achieved over any given period depends on the fraction of the period which is occupied by large amplitude signals, and is thus again dependent on the nature of the input signal.

Transistors can similarly be operated in Class A or Class B. The push–pull Class B transistor amplifier is highly efficient in that during periods of no signal input the total energy dissipated in the stage drops to a very low figure, there being no continuous heater current as in valve amplifiers.

Greater care, however, has to be taken to avoid the introduction of distortion due to "cross-over" effects, as one transistor takes over from the other. In industrial applications, however, distortion is usually of less importance than in radio circuits.

FEEDBACK

Although uncontrolled feedback in amplifiers will produce instability, controlled feedback of the correct phase and amplitude can be used to improve the performance of an amplifier. Its use is very common to improve the response of amplifiers, especially transistor amplifiers. Fig. 13 shows two methods of applying feedback, (a) voltage feedback and (b) current feedback. In the first case, the feedback is proportional to the voltage developed across the load and in the second case it is proportional to the current flowing through the load.

Feedback can be positive or negative. Positive feedback is in phase with the input signal, and is the basis of the oscillator. Negative feedback is in opposite phase to the input signal, and is the most common form of feedback used in amplifiers to stabilise performance.

The effect of negative voltage feedback on the amplifier performance can be summarised as follows: harmonic and intermodulation distortion are reduced; frequency response—

the variation of gain with frequency—is improved and the bandwidth increased; the output resistance of the amplifier is modified; the low frequency characteristics of the output transformer, particularly defects due to the non-linear relation between the flux and magnetising force, are improved;

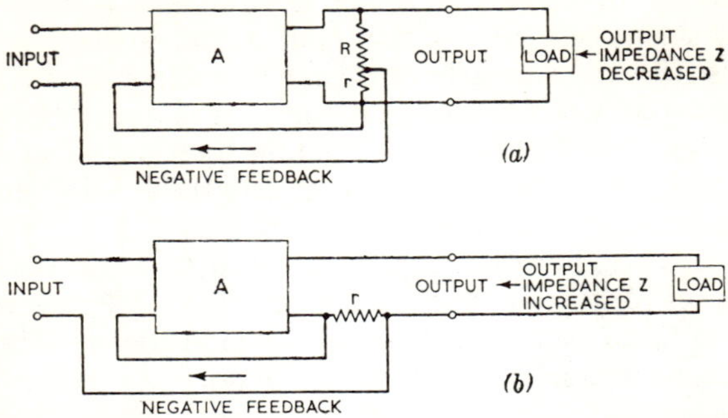

Fig. 13.—*Basic negative feedback circuits.* (*a*) *Voltage feedback.* (*b*) *Current feedback. In these examples the feedback is in series with the input, increasing its impedance. Other circuits may apply feedback to the input in parallel so reducing its impedance.*

the effects of noise or random changes of the parameters of the amplifier originating within the feedback loop are reduced: but all these changes are achieved at the cost of amplifier gain.

Fig. 14 shows the method used in the case of thermionic

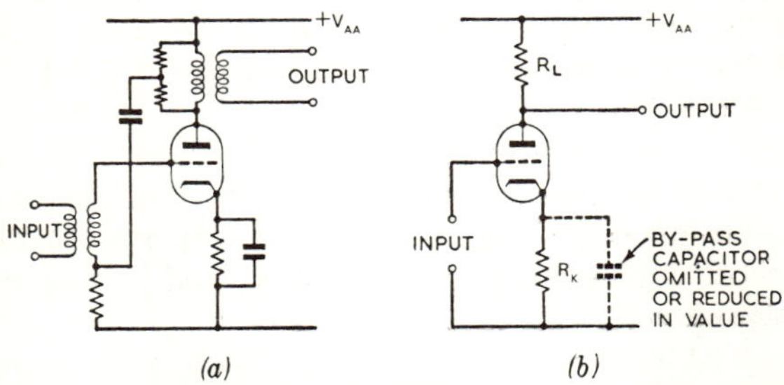

Fig. 14.—*Methods of obtaining feedback in valve amplifier circuits.* (*a*) *Voltage feedback.* (*b*) *Current feedback.*

valve amplifiers to produce (a) voltage and (b) current feedback (i.e. parallel-derived or series derived). When voltage feedback is used the output impedance is *decreased*. In the case of current feedback the output impedance across the anode load is *increased*. Nevertheless, current feedback is often used in practice as it is easiest to apply. The designer, needing improved performance and not wishing to use elaborate circuits, will often leave the cathode bias decoupling capacitors out of the circuit altogether, or at least reduce their value, so that the

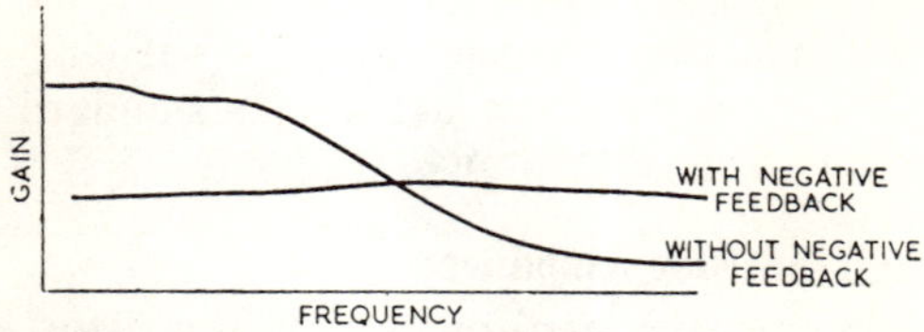

Fig. 15.—*Improvement in a two-stage amplifier obtained by using negative feedback.*

amplification at the higher frequencies is improved relative to that at lower frequencies. Fig. 15 shows the improvement effected in a two-stage thermionic valve amplifier using this method. The input impedance is affected by the manner in which the feedback is applied to the first stage: if in series it increases the impedance and if shunted across the input it decreases the impedance.

The changes of gain, the reduction of distortion or noise, and the increased or decreased input and output impedances of an amplifier are all modified in the same ratio $1+\beta A$, where β is the feedback factor and A the gain of the stage without feedback.

So the gain of a stage with feedback is given by

$$A_f = \frac{A}{1+\beta A}$$

and the increase in output impedance for series-derived feedback would be from Z_o to Z'_o, where $Z'_o = Z_o(1+\beta A)$.

The feedback fraction, β, is the proportion of output signal (voltage or current) that is fed back to the input. For Fig. 13(a) β would be $r/(R+r)$ and for Fig. 13(b) it would be $r/2$.

If the gain of a stage is sufficiently high without feedback the expression for gain with feedback will approximate to

$$A_f \approx \frac{A}{\beta A} = \frac{1}{\beta}$$

so that it will in fact be independent of the valve or transistor parameters—a very important fact when a standard amplifier has to be built to close tolerances.

Feedback in Transistor Amplifiers

Improvement in the performance of transistor amplifiers can also be obtained by the use of feedback. Fig. 16 shows transistor amplifiers in which negative feedback is taken from the output transformer and fed back to the emitter of the transistor. The effect on frequency response and noise is the same as in the valve amplifier. The overall gain is inevitably reduced. The effects of negative feedback on output impedance and input impedance are similar to those for the valve amplifier.

The cathode-follower and transistor emitter-follower stage are both examples of single-stage feedback amplifiers. These are extreme examples where the gain is dramatically reduced and output and input impedances are changed considerably.

In practice, any amplifier may have some degree of negative feedback incorporated. Often it is a mixture of both current and voltage feedback and it brings about an improvement in amplifier characteristics outweighing the resultant loss of gain.

In many circuits it is necessary that the gain shall be constant over a period of time and also between several equipments made from the same circuit drawing. The condition can be achieved by having a carefully controlled negative feedback circuit around a very high gain multi-stage amplifier, as noted

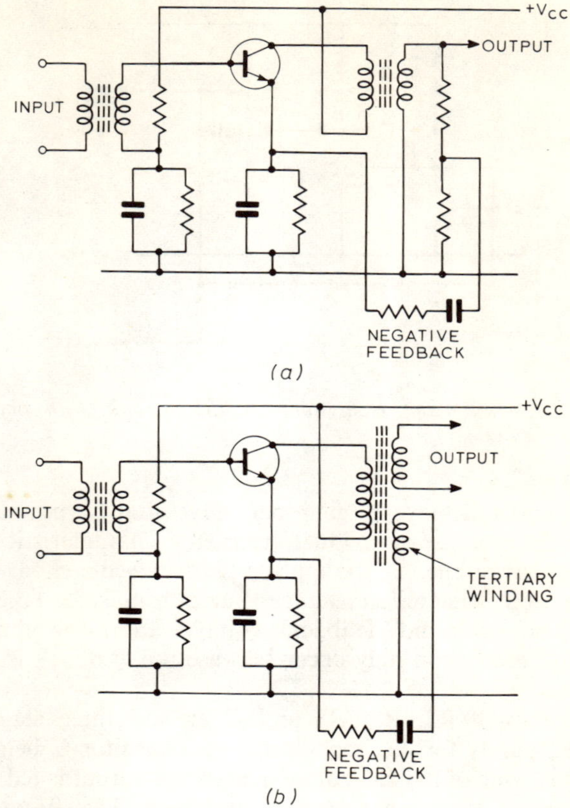

Fig. 16.—*Negative feedback applied to a transformer-coupled transistor amplifier* (*a*) *when output can be earthed on one side*, (*b*) *when output must be balanced and a tertiary winding employed on the transformer to provide the negative feedback voltage*.

earlier and as shown in Fig. 17. The blocks represent d.c. feedback networks, following the principle that any negative feedback should be d.c. wherever possible so that the benefits of better thermal stability may be obtained. At the same time

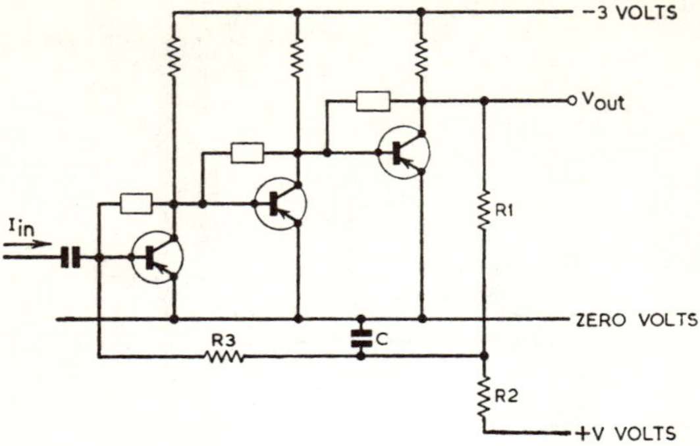

Fig. 17.—*Multi-stage transistor amplifier circuit with negative feedback applied.*

the networks achieve a gain per stage that is practically independent of the individual transistor characteristics. It may be advantageous to apply negative feedback around two stages in some circumstances but care must be taken to ensure that input and "fedback" voltages are not in phase—otherwise oscillation may occur because the feedback is then positive.

The network R1, R2, R3 and C around three stages is provided purely for d.c. stabilising; the capacitor C being so large that none of the a.c. component in the output is fed back to the input at the frequencies to be amplified. The effect of the overall feedback is that the mean of the two voltages $+V$ and V_{out} is fed to the input. If R1 = R2, then $V_{out} = -V$ very approximately. Any drift in the circuit is detected at V_{out} and a correcting signal is applied to the input. Such a circuit can easily have a current gain of 10 per stage, giving an overall current gain of 1,000. The input impedance is about 100 Ω and the output impedance about 2 kΩ, so that the voltage gain is 20,000. The power gain is thus 20×10^6, and the circuit is a very useful a.c. amplifier.

Current feedback in Fig. 18 is also a case worthy of particular note. Here, an undecoupled emitter resistance gives series-derived feedback. This increases both input and output impedances whilst also giving gain reduction. With a high-gain

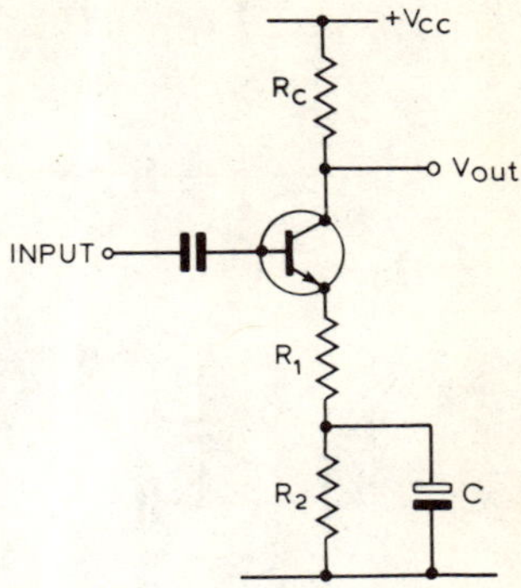

Fig. 18.—*Current feedback using undecoupled emitter resistance.*

transistor in circuit we have already noted that the gain with feedback will be $1/\beta$. In this case R_1 is the series-deriving resistor (assuming C adequately decouples R_2 at the frequencies involved), so $\beta = R_1/R_c$ and the stage gain is then simply $A_f = R_c/R_1$.

SPECIAL CIRCUITS

Phase Splitting

In electronics it is often necessary to have two pulses with opposite polarities yet bearing the same time relationship to the initiating pulse. This is achieved by "phase splitting".

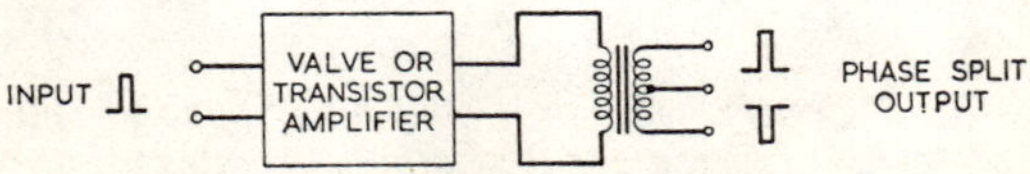

Fig. 19.—*Use of a centre-tapped transformer for phase splitting.*

The two basic methods of doing this are (a) the use of a transformer with a centre-tapped secondary (Fig. 19) or (b) using the 180-degree phase shift that occurs in grounded-cathode and grounded-emitter circuits. In the case of pulses it is difficult to

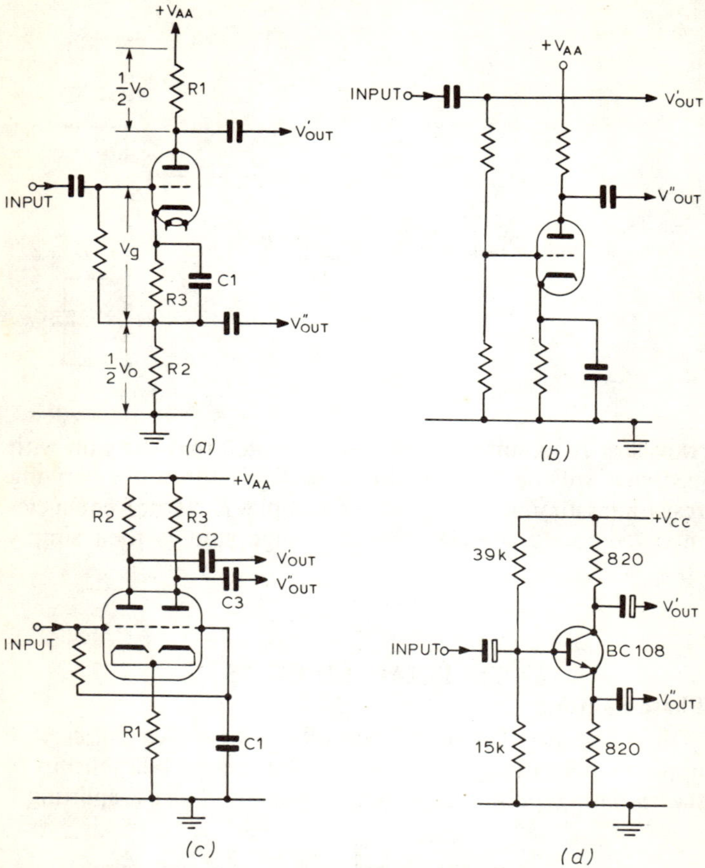

Fig. 20.—(*a*) "*Concertina*" *or* "*split-load*" *phase splitter.* (*b*) "*Paraphase*" *phase reverser.* (*c*) *Cathode-coupled phase splitter.* (*d*) *Transistor version of* (*a*).

design a transformer that will satisfy all the requirements. Pulse transformers with special cores are available, but unless forced to use a transformer by other design requirements, it is simpler to use the alternative method. Three basic valve phase-splitting circuits are shown in Fig. 20. In (a), the load is split between the anode and cathode of the valve and pulses in opposite phase taken from the anode and cathode. In (b), part of the output from the previous stage is applied via a potential divider to the grid of the valve which provides at its anode an oppositely phased pulse: the potential divider is arranged so that the overall gain of the valve path is unity. The arrangement shown in (c), sometimes called a long-tailed pair, is capable of contributing a certain amount of gain to the amplifier. The signal is fed to the grid of the first valve, and applied to the second valve via the common cathode coupling resistor R1. A negative input at V1 grid will provide a positive output at V1 anode and a negative pulse on the cathodes. The negative pulse on V2 cathode is equivalent to a positive pulse on its grid and a negative pulse on its anode. V2 grid is capacitively earthed for cathode input, the resistor connected between the grids ensuring correct d.c. conditions in this section. Transistor equivalents for all these circuits can be made, the comparison for circuit (a) being given at (d).

Transistor "Complementary Symmetry" Circuits

In transistor circuits it is possible to make use of the different characteristics of npn and pnp transistors in a "complementary symmetry" circuit. The main difference

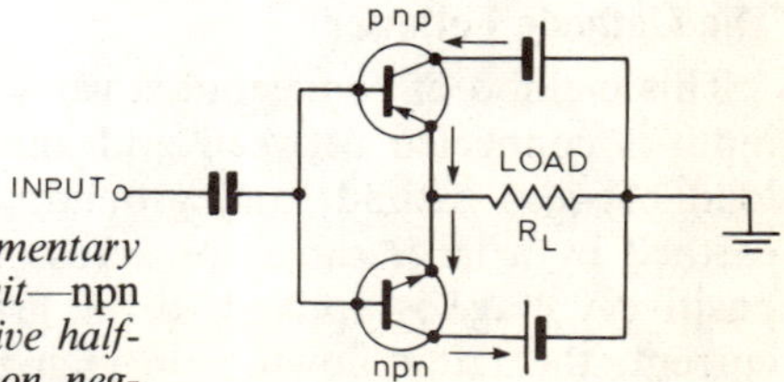

Fig. 21.—*Basic complementary symmetry transistor circuit*—npn *transistor conducts on positive half-cycles and* pnp *transistor on negative half-cycles of the input signal.*

between these two types of transistor is that all polarities required are reversed. A grounded-emitter npn transistor requires a positive drive whilst a grounded-emitter pnp transistor requires a negative drive. If one of each type is connected in parallel the npn transistor will amplify the positive half of an a.c. signal whilst the pnp transistor amplifies the negative half. Thus the collector currents will vary in opposite directions, giving a push–pull output, without the need for a push–pull input. The basic circuit is shown in Fig. 21.

This circuit can be of particular use in audio power amplifiers because no d.c. isolating capacitor is required in series with the load, e.g. a loudspeaker. This can only be achieved when the transistors are operated between two voltage rails—one positive and one negative.

An isolating capacitor would be required if only a single voltage rail were available, with the other rail point at earth potential. In that case the capacitor must have a reactance less than the impedance of the loudspeaker at the lowest frequency that is required; for example, to pass 50 Hz without level reduction into an 8 Ω loudspeaker a capacitor of some 6000 μF would be required!

An important disadvantage of npn/pnp complementary output stages is that for high power working pnp silicon transistors are very much more expensive than the complementary npn; this is because of difficulties in fabrication whilst preserving voltage and current ratings. This means that in commercial designs the advantages of complementary symmetry circuits are outweighed by cost, and Class B stages are often employed.

The Cathode Follower

This method of connecting a valve is shown in Fig. 22. The input is connected between grid and ground and the output load between cathode and ground. The anode is effectively earthed by a large capacitor across the h.t. supply. When a positive voltage is applied to the grid to increase the anode current, the latter flowing through the cathode load R will give a positive change at the cathode; that is the cathode

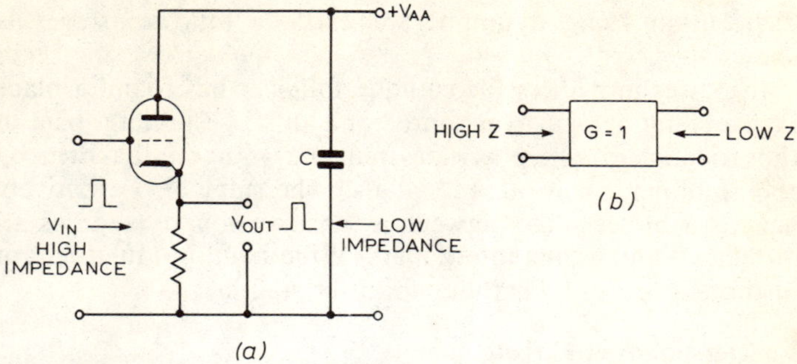

Fig. 22.—*The cathode follower* (*or common-anode amplifier*).

follows the grid in potential, hence the name, cathode follower. This circuit has three noteworthy features: the output impedance is very low, usually less than 300 ohms, whilst the input impedance at the grid is very high—of the order of

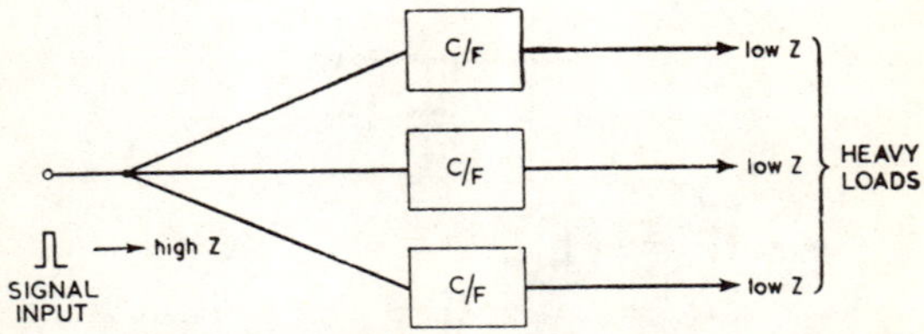

Fig. 23.—*Cathode followers used to distribute outputs.*

megohms. The second feature of importance is that the amplification (gain) of the stage is less than one. That is, no voltage amplification takes place at all. Thirdly, the output has the same phase as the input. The cathode follower is a very useful item of circuitry in the armoury of the electronics engineer. It is often found at the distributing end of some complex equipment where it will accept one signal or pulse, without distorting it (because of the high input impedance), and will then feed into a variety of loads, even one which is changing its value all the time. In Fig. 23 a method of signal distribution using a number of cathode follower stages is shown.

In power amplifiers the cathode follower has found a place in matching a load to a source and this is especially true in those instances where a wide frequency range is handled by the amplifier, as in pulsed signals. The stage also effectively acts as a buffer stage between a sensitive source (such as an oscillator) and a fluctuating load. Wide-band modulators are also often driven by cathode follower stages.

Emitter-Follower Circuit

The transistor equivalent of the cathode follower is the emitter follower, also known as the common-collector or grounded-collector circuit. In this arrangement, the voltage gain is always less than unity. The input is applied between base and collector (see Fig. 24), and the output taken from the emitter circuit. Having high input impedance and low output impedance, this circuit is also useful for matching purposes.

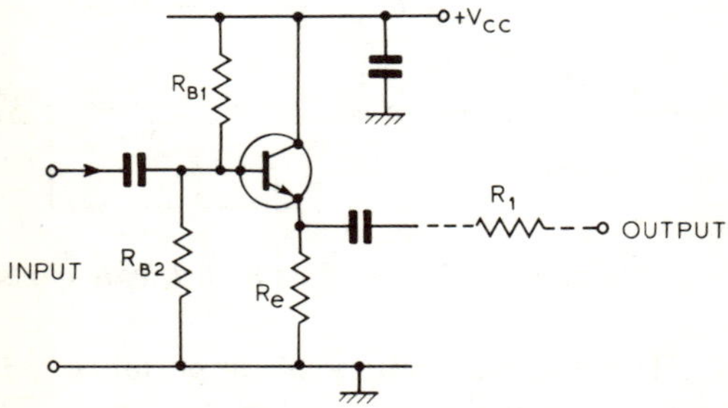

Fig. 24.—*Emitter follower circuit.*

The input impedance of the emitter follower is high compared to a grounded-emitter transistor, the emitter resistor having an effect identical to that of series-applied feedback. A value of input impedance (neglecting the base bias resistors) is $h_{ie}+h_{fe} \cdot R_e$. The output impedance, however, is extremely low and is typically 1–10 Ω. A series resistor R_1 on the output of the stage can therefore be used directly to give the required matching impedance for a particular load.

Cascode Amplifier

There are circumstances when a high-gain, high-input impedance amplifier is required that produces low noise—typically as an r.f. amplifier in a radio or television receiver. A grounded-cathode stage (Fig. 25(a)), if required to produce large gain, is liable to go into oscillation because of anode/grid

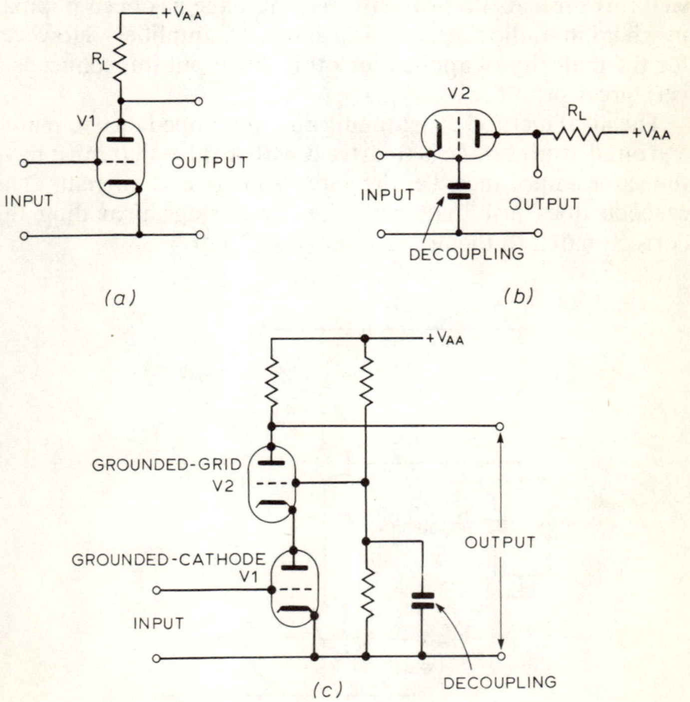

Fig. 25.—*Combination of* (*a*) *grounded-cathode and* (*b*) *grounded-grid stages produces* (*c*) *basic cascode amplifier*.

capacitance feedback (the Miller Capacitance). The stage has, however, the advantage of very high input impedance.

In the grounded-grid stage (Fig. 25(b)) there is no oscillation problem because any anode-grid feedback is decoupled to earth through the capacitor. The stage has the same gain as the common cathode stage because the signal is still applied between grid and cathode. However, in this arrangement the cathode and anode currents are the same and high gain inevitably means high cathode current and therefore very low input impedance. If the signal source has a low impedance as well this presents no problem, and the stage has been popular in v.h.f./f.m. radio receivers as a stable r.f. amplifier. However for the majority of applications this low input impedance is a very great problem.

The ideal method of arranging an input impedance to match a grounded-grid stage is to drive it with a valve that must have the same impedance, i.e. the same voltage and current. The cascode does just that, with the lower stage providing the correct match to the upper stage (Fig. 25(c)).

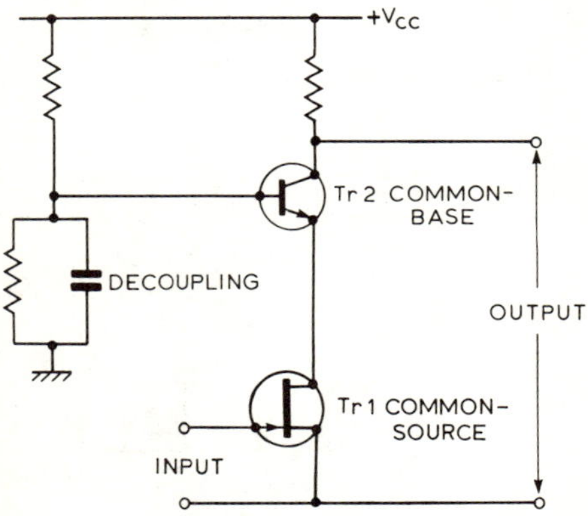

Fig. 26.—*Cascode amplifier using n-channel f.e.t. and* npn *bipolar transistor to give low-noise semiconductor version of Fig. 25.*

The gain of the lower stage is small—because its load impedance (the input impedance of the grounded-grid stage) is very small—so it is stable, yet presents a high input impedance to the source. The gain of the circuit is high (about that of a pentode valve) but the noise is low (about that of a single triode).

The same circuit can be reproduced using bipolar transistors, with the lower stage in common-emitter and the upper stage in common base; but such a circuit would have the grave disadvantage that the input impedance of the grounded-emitter stage would be very low. This can be overcome by using an f.e.t. as the lower element, so that the circuit operates with the same advantages as the valve circuit, with an additional improvement in terms of the low noise of the f.e.t. compared to that of a valve (Fig. 26).

6

OSCILLATORS

We have already noted in Section 5 that feedback, if positive, can cause oscillations to take place. This is the fundamental principle of the operation of any oscillator; for reliable, sustained oscillations to occur the loop gain (βA) must be greater than unity and the loop phase shift must be 360° or its multiple.

SINE WAVE OSCILLATORS

LC Oscillators

The arrangement shown in Fig. 1(a) is typical of a number of commonly used variations in which the feedback principle of oscillation is used with the positive feedback applied via a tuned circuit consisting of inductance and capacitance. The tuned circuit—L1, C1—is connected across the input to the valve. A coil in the anode circuit is inductively coupled to the grid coil. Anode current variations induce e.m.f.s in coil L1: these are arranged to reinforce the e.m.f.s already present in the circuit. If this coupling is sufficient, continuous oscillation is produced.

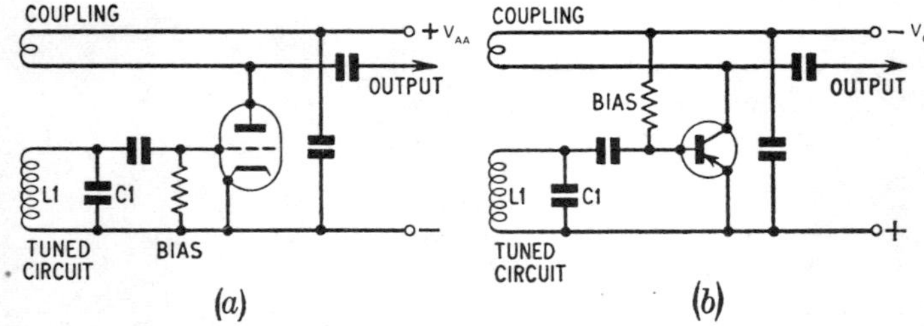

Fig. 1.—(*a*) *Basic valve LC oscillator circuit.* (*b*) *Transistor equivalent —the tuned base LC oscillator.*

The transistor equivalent of the tuned grid oscillator is the tuned base oscillator shown in Fig. 1 (b).

RC Oscillators

In the RC oscillator the feedback is fed from the anode to the grid via a frequency-determining RC network. This type

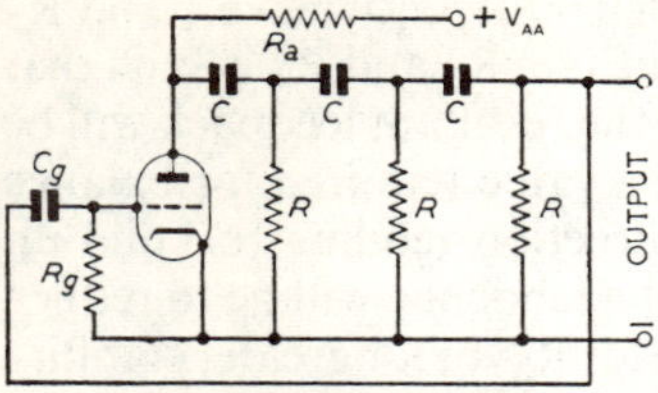

(*a*) RC OSCILLATOR

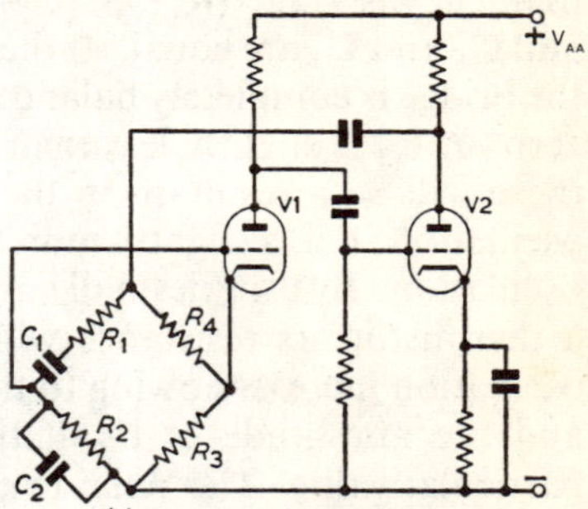

(*b*) WIEN–BRIDGE OSCILLATOR

Fig. 2.—*Two forms of RC phase-shift oscillator. Both circuits can of course use a transistor in place of the valve.*

of oscillator is also known as the phase-shift oscillator. In the example shown in Fig. 2 (a), the feedback is applied via three RC networks in series. Each network introduces a phase shift, the amount depending on the frequency. It can thus be arranged that at one particular frequency the total phase shift is exactly 180 degrees, and at this frequency the circuit will oscillate. The frequency can be changed by varying the values of R or C or both. If $R_a \ll R$ and all capacitors C and resistors R are alike, the oscillator frequency is $1/2\pi RC\sqrt{6}$, and for oscillations to start the amplifier must have a voltage gain of 29. For continuous frequency variation a three-gang variable capacitor could be used, provided all rotor sections are not commoned.

The RC oscillator can produce a very pure output over a frequency range from a fraction of a hertz to several hundred kilohertz.

A further form of RC oscillator is the Wien-bridge oscillator, the circuit of which is shown in Fig. 2 (b). This circuit employs two valves or transistors and the feedback is applied via a bridge circuit. The arrangement has good frequency and amplitude stability. Feedback from V2 anode is applied across two corners of the bridge to V1 grid. V1 grid is excited by the voltage existing across the other two corners of the bridge. As there are two valves the required condition for sustained oscillation is that the feedback is in phase. This occurs when the frequency is $1/2\pi\sqrt{(R_1R_2C_1C_2)}$ or $1/2\pi R_1C_1$ when R_1 and R_2 and C_1 and C_2 are equal. If the ratio of R_3 and R_4 is such that the bridge is completely balanced the resultant feedback will be zero. If R_3 is slightly less than the value required for balance there will be a resultant in the direction tending to build up oscillation: if it is slightly more, the tendency will be to reduce oscillation. By using a small lamp for R_3, or for greater stability a thermistor, its resistance will increase as the amplitude of oscillation increases owing to the increased current through it, and the amplitude of oscillation will thus be limited to a particular value. The Wien-bridge oscillator is primarily a low frequency oscillator having precision and a pure waveform. Because of the need for slight unbalance in the bridge to provide sufficient feedback to make up for losses, the working frequency in practice is very slightly different from the theoretical value.

Beat-frequency Oscillators

The beat-frequency oscillator covers a wide frequency range in one sweep. It employs the beats between two r.f. oscillators, one fixed and the other variable. An oscillator variable from 100 to 110 kHz, for example, beating with a fixed oscillator at 100 kHz will produce an a.f. output varying from 10 kHz down to nearly zero. In practice, interaction occurs when the two oscillator frequencies are very close, so that the lower limit is about 20 Hz. The waveform of a beat-frequency oscillator is not as good as that of an RC oscillator, and small changes in the frequency of the r.f. oscillators can produce large changes in the beat frequency so that frequent resetting may be necessary.

Signal Generators

A signal generator consists in essence of an oscillator covering a certain predetermined frequency range. The

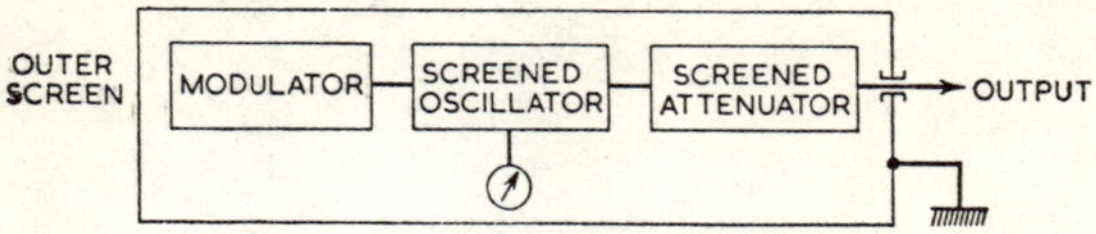

Fig. 3.—*Basic elements of a signal generator.*

oscillator must be carefully screened in order to avoid stray radiation. The output is fed to the output terminals via a calibrated attenuator. The essential components of a signal generator are shown in Fig. 3.

Crystal Oscillators

The application of a voltage across certain types of crystal produces a mechanical deformation. This phenomenon can be used to obtain a mechanical resonant element which has frequency stability of a high order. Fig. 4 (a) shows the electrical equivalent circuit of a crystal. L, C and R correspond to mechanical inertia, compliance and damping respectively; C_0 is the capacitance between the electrodes applied to the crystal, and is due to the mounting, the holder and the wiring of the crystal into the circuit. L and C are very much greater and smaller, respectively, than the corresponding quantities in a normal electrical circuit, and the Q-factor is very much higher, that is the damping is very low and consequently the tuning is very sharp. The crystal has a series resonant frequency at which the reactances of L and C are equal and opposite. At a slightly higher frequency, when the crystal itself behaves as an inductance, a parallel resonance occurs between this resultant inductance and the capacitance C_0. These two frequencies are very close together, generally well within one per cent.

The crystal may be connected to a valve or transistor oscillator to maintain continuous oscillation. The two basic

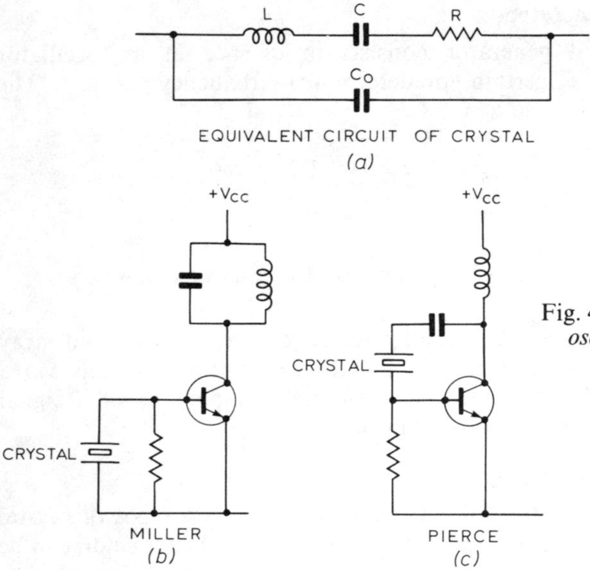

Fig. 4.—*Crystal oscillators.*

transistor circuits are shown in Fig. 4(b) and (c). Feedback occurs through stray capacitance in the valve. In the Miller oscillator the collector circuit should be tuned to a frequency slightly above the resonant frequency of the crystal, which then oscillates at a frequency just below the parallel resonance point.

A frequency stability of one part in a few million is easily obtained with a crystal-controlled oscillator. If the crystal is maintained in a temperature-controlled chamber the stability can be better than one part in 10^8, and certain "sub-standard" transmitters use complex stabilisation techniques to achieve guaranteed stabilities of better than one part in 10^{10}.

SELF-OSCILLATING PULSE GENERATORS

The Multivibrator

Fig. 5 shows a transistorised self-oscillating multivibrator. Such circuits are often called free-running or astable multivibrators to distinguish them from the triggered flip-flop and bi-stable types of multivibrator. In operation, each transistor is automatically switched from a non-conducting state to a fully conductive (bottomed) state by feedback from the collector circuit of the other transistor.

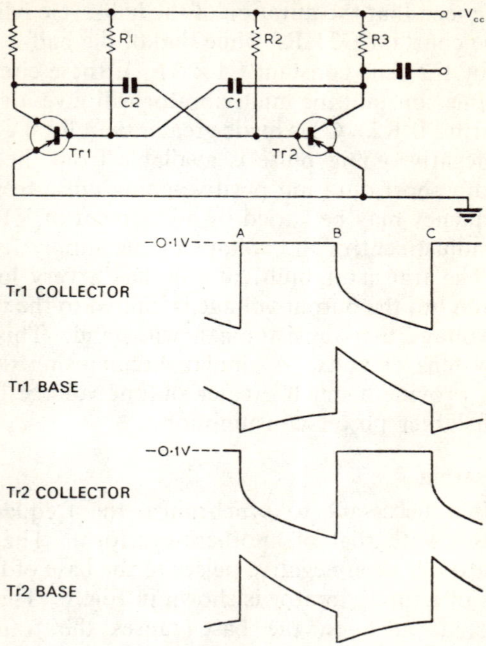

Fig. 5.—*Transistor multivibrator.*

When Tr1 becomes conductive at time A, its collector voltage changes from the value of the supply voltage V_b to about −0·1 V. This voltage change is passed via capacitor C2 to the base of Tr2, cutting off the collector current of Tr2. The circuit remains in this state while capacitor C2 discharges via R2. Meanwhile, C1 charges rapidly via R3 and the base current of Tr1. When C2 has discharged sufficiently to allow Tr2 to pass collector current (time B), the collector voltage of

Tr2 changes towards $-0{\cdot}1$ V, the change being passed by C1 to cut off Tr1, hence the collector voltage of Tr1 becomes more negative with the result that Tr2 is driven into its bottomed state. C1 now discharges via R1 until time C, when Tr1 again becomes conductive and the next cycle starts.

It will be seen that thc duration of the half-cycle AB depends on the time constant C2 × R2 while that of the half-cycle BC is governed by the time constant C1 × R1. If these circuits have the same time constant the multivibrator will give a symmetrical waveform. If R2 × C2 is made greater than R1 × C1, a short duration negative-going pulse is available from the collector of Tr2 and a short duration positive-going pulse from Tr1.

The frequency may be varied by adjustment of R1, R2, C1, C2 or by adjustment of the voltage of the supply feeding R1 and R2. The transistor multivibrator has a very low power consumption but the output voltage is limited to the maximum collector voltage the transistor can withstand. This is often only a few tens of volts. A similar circuit using thermionic valves can provide a much greater output voltage but has a very much larger power consumption.

Synchronisation

It is often necessary to synchronise the frequency of a multivibrator with that of another waveform. The effect of feeding short duration negative pulses to the base of one of the transistors of a multivibrator is shown in Fig. 6. The negative synchronising pulse on the base causes the transistor to become conductive earlier than when running freely. Thus the frequency of waveform (c) is higher than the free-running frequency (a).

Frequency Division

Waveforms (d) and (e) of Fig. 6 show the effect of synchronising a multivibrator by a frequency slightly greater than ten times the free-running frequency. The two transistors now reverse their conductive states every fifth synchronising pulse, hence the frequency of the output waveform is accurately

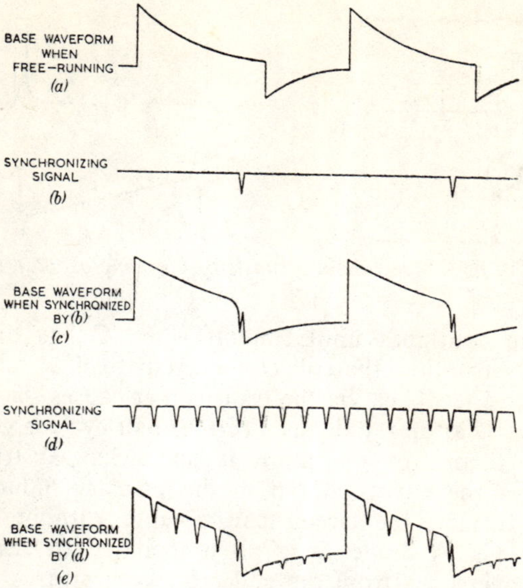

Fig. 6.—*Synchronising a multivibrator.*

locked to one tenth of the frequency of the synchronising pulses. If the output is differentiated it can be used to synchronise another multivibrator, the frequency of which is slightly less than one-tenth of the first multivibrator, giving a combined circuit capable of dividing the frequency of the original input by one hundred.

The Free-running Blocking Oscillator

Fig. 7 shows a transistor self-oscillating blocking oscillator. The circuit is similar to that of a sine wave oscillator in which the inductance and capacitance of the transformer form a resonant circuit. The coupling between primary and secondary windings is, however, so tight that the transistor becomes biased off after one half-cycle, capacitor C becoming rapidly charged so that the collector current is quenched or blocked at time A.

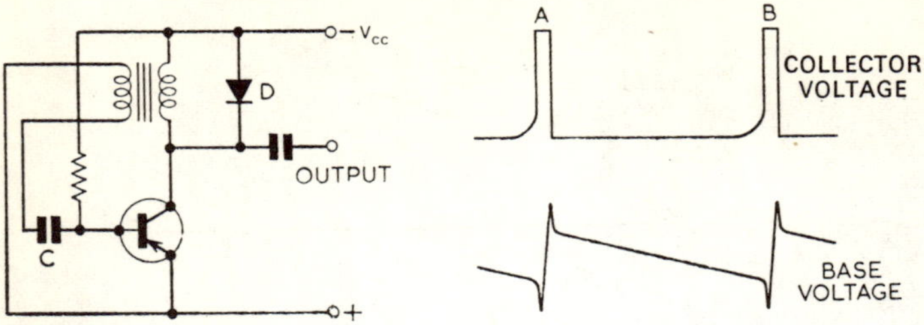

Fig. 7.—*Free-running transistor blocking oscillator.*

This state continues until time B when C has discharged sufficiently to allow the collector current to flow once more. The positive feedback via the transformer causes the collector current to build up rapidly as a further half-cycle is generated. The full theory of operation is somewhat abstruse and, because of the energy stored in the primary inductance of the transformer, the circuit can operate without the base capacitor C. A diode D is often included to remove the negative overshoot from the collector waveform.

Complementary Transistor Square Wave Generators

A pnp and an npn transistor may be combined as shown in Fig. 8 to form a bi-stable switch. Depending on the emitter/ base voltages, the transistors are either both conductive or

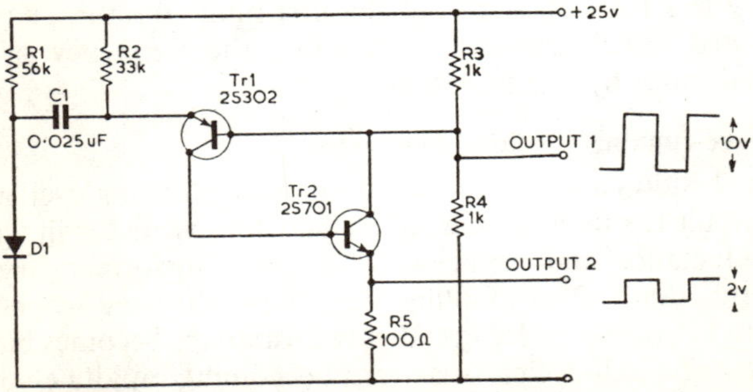

Fig. 8.—*Complementary transistor bi-stable switch.*

both non-conductive. The inclusion of capacitor C1 converts the circuit into a self-oscillating square wave generator. Initially, the emitter/base junction of Tr1 is reverse biased so that both transistors are non-conductive. Capacitor C1 charges via diode D1 and resistor R2. When capacitor C1 has charged sufficiently to raise the emitter voltage of Tr1 above that of its base, both transistors become conductive giving a rapid change in voltage at the collector of Tr2. Both transistors are retained in this state by regenerative feedback to the base of Tr1. This state continues until capacitor C1 has discharged via R1 and the transistors sufficiently to cause the emitter voltage of Tr1 to fall below the new voltage at the base, when both transistors become non-conductive and the cycle repeats.

Short rise and decay times can be obtained with this circuit since the regenerative feedback loops contain no attenuating components other than the transistors themselves. The circuit shown in Fig. 8 is a basic 1 kHz square wave generator providing either a relatively large voltage output or heavy current pulses. The frequency is largely independent of changes in supply voltage. The mark/space ratio may be adjusted by altering the relative values of resistors R1 and R2.

SINGLE-STROKE PULSE GENERATORS

If one side of a multivibrator is biased into a non-conductive state, the circuit will remain in this state until a trigger pulse is received. It will then undergo a single cycle of oscillation before returning to its stable state. These circuits are often called monostable multivibrators, single-shot multivibrators, monostable trigger circuits or flip-flops. The last term is, however, used in the U.S.A. for bi-stable circuits, i.e. circuits which change their conductive state on receipt of a trigger pulse and remain in that state until the arrival of the next trigger pulse.

Fig. 9 shows a transistorised flip-flop. This circuit is similar to the self-oscillating multivibrator shown in Fig. 5 but the capacitor coupling the collector of Tr2 to the base of

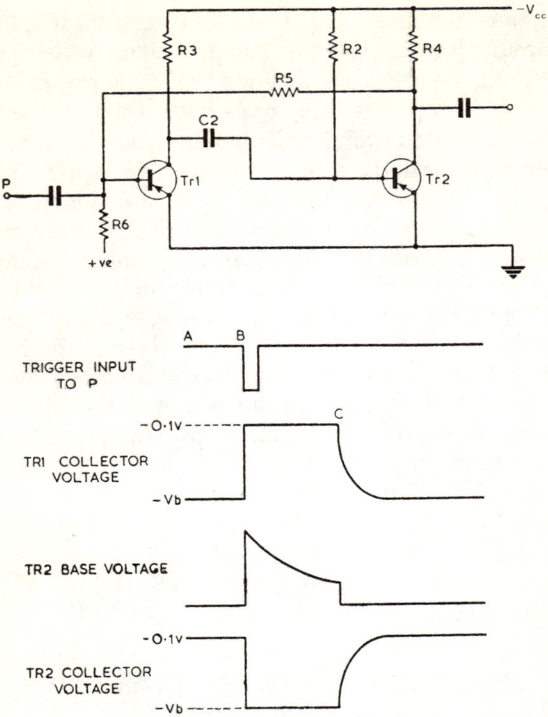

Fig. 9.—*Transistor single-stroke multivibrator or flip-flop.*

Tr1 is replaced by a resistor R5, and the base leak R6 is fed from a positive potential +ve sufficient to prevent Tr1 passing collector current. Tr2 is biased into the conductive state by the negative supply $-V_b$ feeding R2. Hence, in this stable state, shown at time A, Tr1 is cut off so that its collector voltage is at $-V_b$; Tr2 is bottomed hence its collector voltage is approximately $-0{\cdot}1$ V. Let us assume a trigger pulse arrives at terminal P at time B. Tr1 will be driven into conduction, hence the collector voltage will rise towards $-0{\cdot}1$ V. This change is passed by C2 to cut off Tr2. The change in collector voltage of Tr2 will be passed via R5 to maintain this transistor conductive at the end of the trigger pulse. This condition is called the unstable state and is maintained until C2 has discharged via R2 sufficiently to allow Tr2 to conduct again. This is shown at C. The voltage step at the collector of Tr2 is passed to the base of Tr1 to return this transistor to its stable state.

C2 now recharges via R3 and the base/emitter path of Tr2. Each time a trigger pulse is received, the circuit generates a positive-going pulse at the collector of Tr1 and a negative-going pulse at the collector of Tr2. The pulse duration is independent of the duration of the trigger pulse and is controlled by adjusting the value of C2, R2 or the negative voltage feeding R2. The circuit is thus distinguished from the synchronised self-oscillating multivibrator which continues oscillating at a slightly different frequency in the absence of input pulses.

The Triggered Blocking Oscillator

If the base leak of a blocking oscillator is fed from a positive bias supply, the transistor will be non-conductive until a negative trigger pulse is received at the base, when it will generate a single half-cycle of oscillation. The biasing circuit will then take control so that the circuit remains inoperative until the next pulse is received.

SAWTOOTH VOLTAGE GENERATORS

The most important use for a sawtooth waveform is the provision of a timebase voltage for a cathode-ray oscilloscope; sawtooth waveforms are also used in pulse-delaying circuits and in the generation of pulses of controlled duration. Most sawtooth generators use the principle of charging a capacitor comparatively slowly and discharging it more rapidly via some form of electronic switch.

The Thyratron Timebase

The build-up of voltage across capacitor C—Fig. 10 (a)—follows an exponential curve E, the form of which is dependent on the time constant C × R. When the breakdown voltage of the gas-filled triode or thyratron T is reached, the capacitor C is partially discharged by the heavy anode current, as shown at F. The anode voltage now drops rapidly until the thyratron-sustaining voltage is reached, when (provided the total anode circuit resistance is sufficient) the valve again behaves as an open circuit. The next cycle now begins, the capacitor C

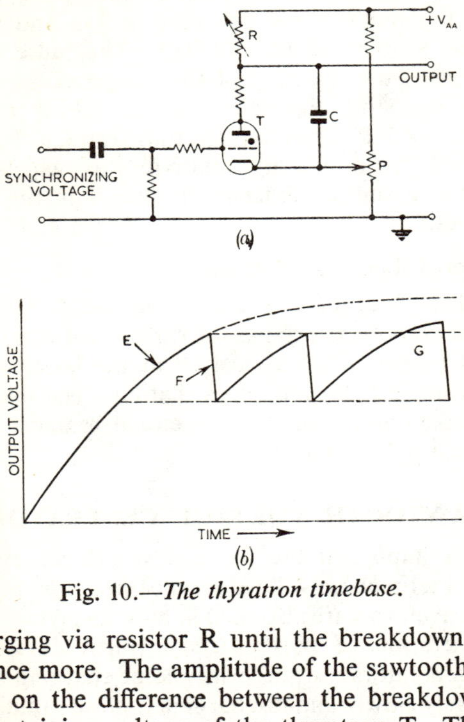

Fig. 10.—*The thyratron timebase.*

again charging via resistor R until the breakdown voltage is reached once more. The amplitude of the sawtooth voltage is dependent on the difference between the breakdown voltage and the sustaining voltage of the thyratron T. The latter is governed by the construction of the valve, but the former may be controlled by adjustment of the steady voltage applied to the grid or cathode by the potentiometer P. The effect of increasing this voltage is to increase the amplitude of the sawtooth waveform, as shown at G. This control also has some effect on the frequency. The main frequency control is, however, obtained by adjustment of the charging resistor R, a coarse adjustment being obtained by switching in different value capacitors.

In an oscilloscope, the part of the wave being examined must be fed to the grid to synchronise the sawtooth frequency to a sub-multiple of that of the examined waveform. The frequency controls are adjusted so that in the absence of synchronising signals the frequency would be slightly too low. The thyratron is made conductive at the right instant by a positive half-cycle of the synchronising voltage.

Linear Timebase Generators

The rate of increase of voltage when a capacitor is charged via a resistor decreases as the capacitor becomes charged. This results in a pronounced curvature of the upper part of the sawtooth waveform shown in Fig. 10 (b). It will be noticed that the rise at the foot of each cycle is very nearly linear. If the capacitor were charged through a much higher resistance from a very high voltage source, but discharged after it had reached only a small proportion of the charging voltage, a much more linear sawtooth wave would result. The linearity and voltage output are inversely proportional to one another.

Many linear timebase circuits have been devised which avoid the need for a very high voltage supply. The charging resistor may be replaced by a pentode or a transistor or even a triode with a high value un-bypassed cathode resistor. These devices have the property of passing a constant current substantially independent of the voltage across them. If the current is constant, the rate of charge is constant, and hence the voltage rise is linear.

The "Bootstrap" Timebase

Another circuit for generating a linear sawtooth waveform is shown in Fig. 11. Capacitor C1 is charged from the negative supply V_b via R1 and a diode D. At the end of each sweep, C1 is rapidly discharged by means of an electronic switch represented by contact S. The rise in voltage across C1 is passed by an emitter-follower transistor T to the upper end of the charging resistor R1. Since the emitter follower gain is almost equal to one, the voltage across R1 is maintained nearly constant whilst C1 is being charged, hence the charging current is almost constant. The voltage across C1 is used to control the voltage at the emitter, which in turn is used to increase the rate of rise of the capacitor voltage so as to improve its linearity. This circuit is often likened to a man attempting to lift himself by his own bootstraps. The circuit is, however, far more successful than its human analogy.

The emitter follower can of course be replaced by a valve connected as a cathode follower.

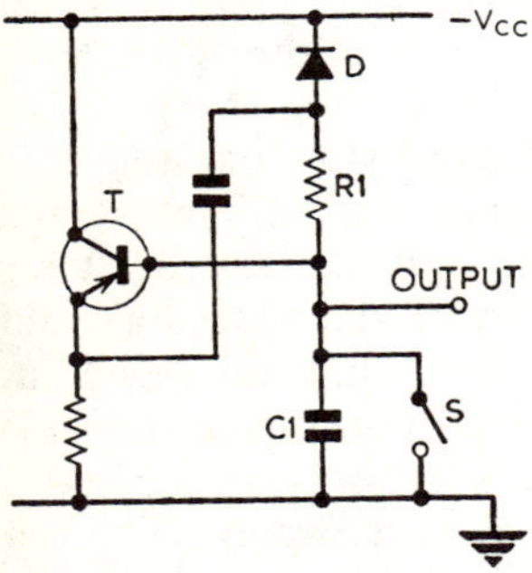

Fig. 11.—*The bootstrap timebase.*

The Miller Timebase

If a Miller integrator is fed from a direct voltage supply and the amplifier is alternately switched on and off, a linear sawtooth waveform will be generated. Fig. 12 shows a basic form of Miller timebase. The anode current is controlled by a switching waveform fed to the suppressor grid. During negative half-cycles of the switching waveform (time A) the anode current is cut off, hence the voltage rise at the anode will drive the grid positive causing grid current to charge

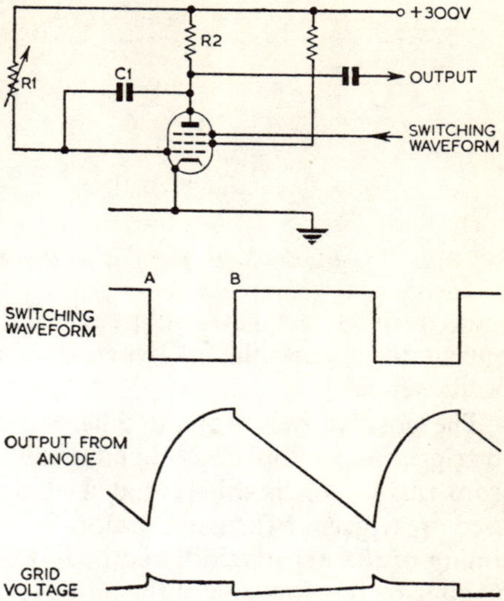

Fig. 12.—*The Miller timebase.*

capacitor C1 via resistor R2. At time B, the switching waveform allows anode current to flow and capacitor C1 discharges slowly via R1, R2 and the valve. As the discharge current tends to decrease, the grid voltage tends to rise so that the anode voltage tends to fall more linearly. The switching waveform may be derived from a self-oscillating multivibrator so that the complete circuit is free-running.

WIDE-RANGE PULSE GENERATORS

Many commercial pulse generators provide wide variations of pulse repetition frequency (p.r.f.) and pulse duration (pulse width). Fig. 13 shows a typical instrument. The p.r.f. is determined by a variable frequency oscillator 1 which may be a sine-wave oscillator or a self-oscillating multivibrator. The output is shaped by an overdriven amplifier (squarer) 2 to form a variable frequency square-wave output at 3. The square wave is differentiated at 4, the positive pulses being

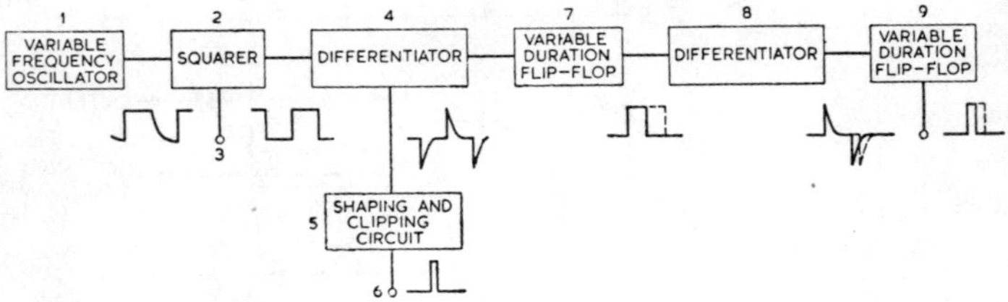

Fig. 13.—*Block diagram of a wide-range pulse generator.*

shaped and the negative clipped at 5 to provide a prepulse output at 6, suitable for synchronising the timebase of an oscilloscope.

The positive pulses of the differentiated waveform are used to trigger a flip-flop 7 having adjustable duration. The output from this circuit is differentiated at 8 and the negative pulse used to trigger a further flip-flop 9. It will be seen that the timing of the negative differentiated pulse from 8 is adjustable by altering the duration of the output pulse from the flip-flop 7. The main output from 9 thus occurs after the pre-pulse from 6 at a time determined by the setting of the flip-flop 7. The duration of the main output pulse is governed by the flip-flop 9.

Generators offering very fast rise times and/or short pulse widths may use different pulse-generating techniques, but the principles are similar.

Silicon Controlled Switch (S.C.S.) Oscillator

In some applications such as the field timebase of television receivers—operating at 50 Hz—the sinewave oscillator is impractical because of the large inductor that would be required. The blocking oscillator is practical (the transformer in the blocking oscillator is used as such and not used as an inductor), and the multivibrator or self-oscillating multivibrator is practical. Because receiver manufacturers prefer, if possible, to avoid the use of wound components due to their relatively high cost and the expense of having them set to their correct tuning position on the production line, the majority of earlier receivers used the multivibrator or the self-oscillating multivibrator. However, the transistor is not very temperature stable in the multivibrator connection and should therefore be avoided.

If the receiver is fully solid-state, one way of avoiding these problems is by the use of an oscillator employing a silicon controlled switch. The s.c.s. is a four-layer semiconductor device similar to the thyristor but with all four layers brought out to terminal connections. Fig. 14 shows the layer arrangement and the symbol of nomenclature used with the device.

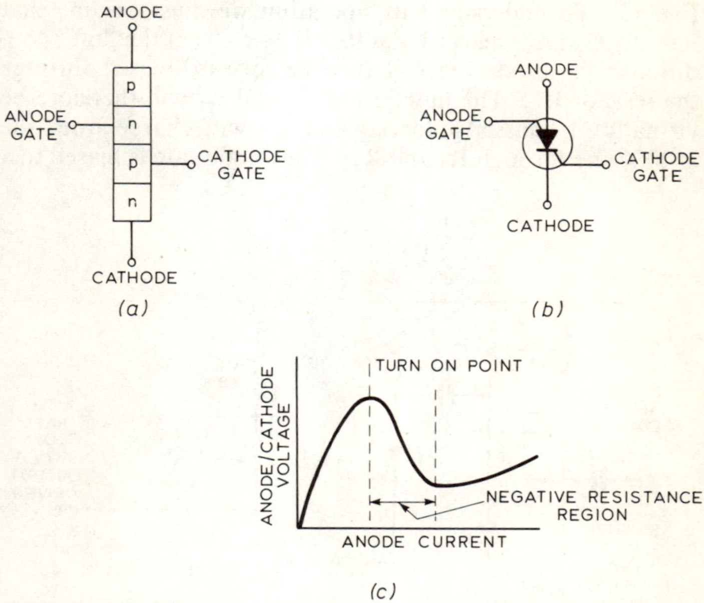

Fig. 14.—*Silicon controlled switch:* (*a*) *layer structure;* (*b*) *symbol;* (*c*) *plot of anode/cathode voltage versus anode current.*

With the two gates left unconnected (i.e. zero voltage) a very small leakage current will flow with a positive anode/cathode voltage. As this voltage is increased a breakdown point is reached where the device enters a negative-resistance region (similar to the tetrode valve)—Fig. 14(c). This is a low impedance condition. The device will continue this conduction for as long as the anode current is allowed to remain above a "holding current" level. Below that level it returns to being a high impedance device between anode and cathode.

With the two gates connected to suitable supplies, the s.c.s. can be made to turn on below its normal threshold with forward biasing of the intermediate junctions in the device.

A representative circuit for an s.c.s. field oscillator is given in Fig. 15. To understand its operation we must assume that initially C_1 is discharged and that the s.c.s. is in the "off" condition. The diode D_1 will then be forward-biased through the resistor R_1. The junction of C_1 and R_1 will therefore be virtually at chassis potential and C_1 will charge from the +40 V line through R_2 and R_3. The anode gate is biased to a

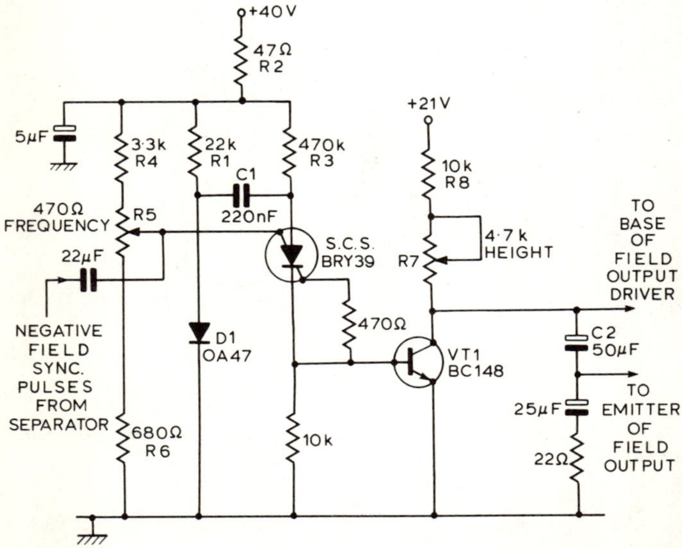

Fig. 15.—*S.C.S. oscillator for field timebase* (*Philips and Bush*).

positive potential set by the potential divider chain R_4, R_6 and the frequency control R_5. When the anode voltage is just greater than the anode gate (which will occur while the capacitor C_1 is charging) current will flow in the s.c.s. turning it "on".

This low-impedance state passes current derived from the anode gate potential divider and the discharge current of C_1 and the increase in cathode voltage turns on the transistor VT1. Across VT1 is the charging capacitor for the timebase, C_2, which is charged from the 21 V rail through the height control R_7 and R_8. The charging and discharging of C_2 is therefore controlled by the switching on and off of VT1.

As the capacitor C_1 is discharging through the s.c.s. there comes a point where the voltage on the R_1 side of the capacitor is greater than that on the R_3 side (R_3 is larger than R_1). At that point the diode D1 suddenly conducts and the cycle is completed as the s.c.s. falls below the holding current and turns off. The scanning period time is dependent on the values of R_3 and C_1 whilst the flyback time is dependent on the values of R_1 and C_1. Both times are also dependent on the setting of the frequency control, R_5, under the free-running conditions. When synchronised by negative pulses appearing on the anode gate, R_5 has no control.

7
D.C. AMPLIFIERS

Coupling

The use of capacitors or transformers for coupling amplifier stages blocks the d.c. component on which a.c. signals are superimposed and enables the biasing circuit for each stage of an amplifier to be designed independently of the others. These a.c. coupling techniques obviously cannot be used in circuits designed for the amplification of d.c. or very slowly varying signals, so techniques for d.c. coupling have been devised. Such techniques do not preclude the use of the circuits for a.c. amplification and they are often used for this purpose in integrated circuits because of their economy of slice area.

Drift

The performance of a transistor may alter in respect to current gain, leakage current and base–emitter voltage due to thermal effects. A consequence of this may be that the amplifier output voltage will change without any change of input voltage, an effect called *drift*. Drift cannot be distinguished from the output changes which are required due to input changes, so it must be kept to a minimum. Methods of minimising drift include the use of feedback, heat sensitive elements to produce compensating signals, and chopper amplifiers in which the signal is converted to a square wave, amplified in an a.c. amplifier and converted back to d.c.

Darlington Pair

A *Darlington* or *super alpha pair* is shown in Fig. 1. It is sometimes called a composite transistor, and indeed two (or even three) transistors may be encapsulated into a simple package with just three output leads. The main advantage of

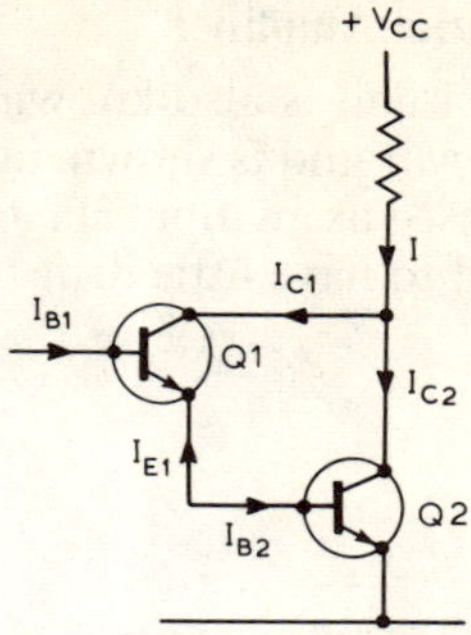

Fig. 1.—*Darlington pair.*

this circuit is a higher input impedance than would be achieved with a simple stage. If used as in Fig. 1, Q_1 is an emitter follower stage and Q_2 a common emitter amplifier, so the voltage gain is no higher than for a single stage. The overall current gain I/I_{B1} may be found as follows:

$$I_{E1} \approx -(1+h_{FE1})I_{B1}$$
$$\approx -h_{FE1} \times I_{B1} \text{ (if } h_{FE1} \text{ is large)}$$

Now the emitter current of Q_1 is the base current for Q_2 and, using the inward flowing current sign convention,

$$I_{B2} = -I_{E1} = h_{FE1} \times I_{B1}$$
$$I_{C2} = h_{FE2} \times I_{B2}$$
$$I = I_{C1} + I_{C2}$$

Again if h_{FE2} is large I_{C1} may be ignored and

$$\frac{I}{I_{B2}} = h_{FE1} \times h_{FE2}$$

As regards drift the Darlington connection is not ideal since the leakage current of Q_1 is amplified by Q_2. Very high values of input resistance can be obtained by connecting the load resistance into the emitter circuit.

Differential Amplifier

This circuit is also known as the *long tailed pair* or *emitter coupled pair* and is shown, in its balanced form, in Fig. 2. It is widely used as an input stage for a d.c. amplifier since it can be designed to have little drift.

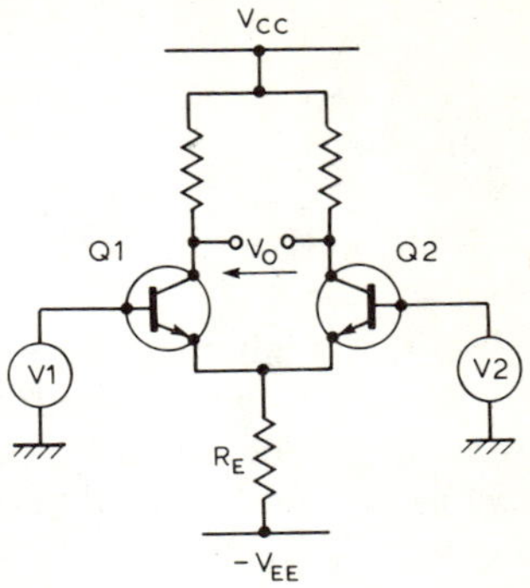

Fig. 2.—*Basic differential amplifier*.

To begin with, consider the case where the two input voltages V_1 and V_2 are equal in amplitude and phase. Then although the collector voltages will change they will change together, and if the two sides of the amplifier are identical then $V_0 = 0$ V. Such input voltages are called common mode. On the other hand if the input voltages are not the same (for example with a d.c. input), one collector current will change by a different amount from the other and V_0 will be the difference between the two collector voltages; it will be an amplified version of the difference between the input voltages.

Apart from differences between the two sides of the amplifier, the current may produce an output voltage for common mode signals because of thermal effects or non-linearity. Ideally, the output voltage for $V_1 = 0{\cdot}1$ mV, $V_2 = -0{\cdot}1$ mV (case A) should be the same as for $V_1 = 1{\cdot}1$ mV, $V_2 = 0{\cdot}9$ mV (case B). Suppose that in case A $V_0 = 40$ mV and in case B $V_0 = 42$ mV. The difference input in each case is 0·2 mV and produces an output of 40 mV giving a difference gain $A_d = 200$. The common mode signal in case B is 1 mV and produces an additional output of 2 mV, so the common mode gain $A_c = 2$. The figure of merit here is called the Common Mode Rejection Ratio (CMRR), and is defined as CMRR $= A_d/A_c$, which in this example will be 100.

The CMRR can be increased by increasing the value of R_E since any common mode signal produces a larger feedback voltage across it but there are practical limitations set by the bias voltage required across R_E. Since operational considerations will fix the quiescent values of collector voltage, the collector currents will be fixed, so that increasing the value of R_E will require an increase in the bias supplies. However it is worth pointing out that R_E does not cause any loss of gain due to feedback, because as one collector current rises to produce a negative feedback voltage across R_E the other falls to offset this.

A practical circuit which gives a very large effective value of R_E while avoiding the need for a large bias voltage is to use a transistor to determine the emitter currents as shown in Fig. 3. The total emitter current, I_E, will be determined by

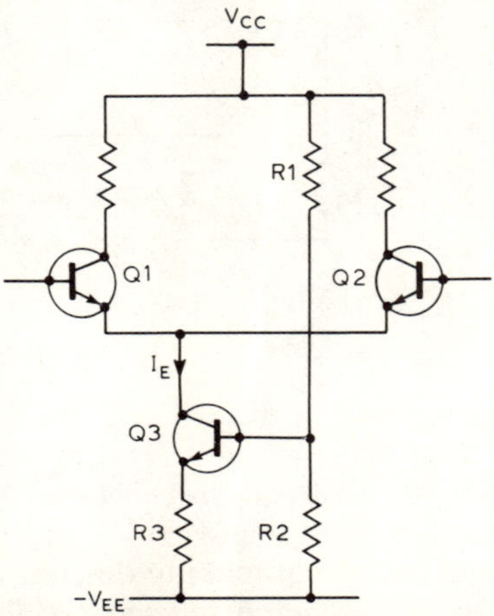

Fig. 3.—*Using a transistor to determine* I_E.

R_1, R_2 and R_3 and since the resistance looking between C and E of Q_3 is very high then so will be the effective value of R_E.

Unbalanced Differential Amplifier

There is often a requirement for single-ended input and output voltages, and suitable circuits for this purpose are shown in Fig. 4 (a) and (b). In Fig. 4 (b) the circuit is essentially an emitter follower Q_1, feeding a common base amplifier Q_2.

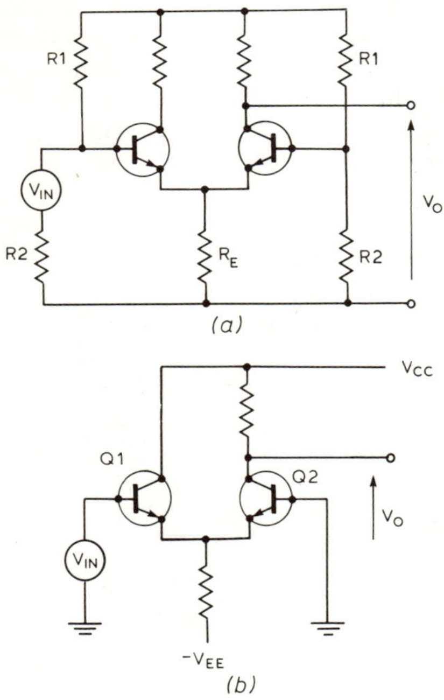

Fig. 4.—*Unbalanced differential amplifiers.*

A particular advantage of making differential amplifiers in integrated circuits is that manufacturers are able to match the transistor parameters accurately. Also, any parameter changes due to thermal effects will be common to both transistors and will not produce any drift.

Operational Amplifiers

An operational amplifier is one which carries out, in analogue form, certain mathematical operations such as addition, differentiation and integration. The type of operation to be carried out is determined by a feedback network in an amplifier of high gain and low drift.

Suppose we have an amplifier of large negative voltage gain $-A$ as shown in Fig. 5. Then as V_1 goes positive V_2 goes

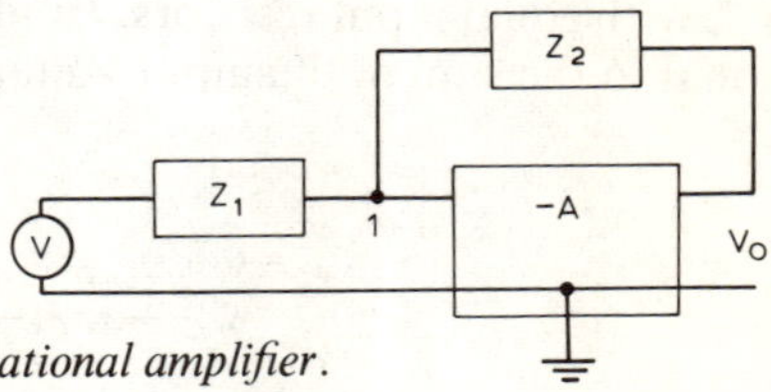

Fig. 5.—*Operational amplifier*.

negative and, because of Miller effect, V_2 tends to hold terminal 1 at earth potential. Terminal 1 is said to be a virtual earth. In this situation $V = IZ_1$. If the amplifier itself has a high input impedance the current in Z_2 will also be I and $V_0 = -IZ_2$. Then the overall gain A_0 will be

$$A_0 = \frac{V_0}{V} = \frac{-Z_2}{Z_1}$$

In other words the gain, i.e. the relationship between input and output voltages, is solely determined by the external components Z_1 and Z_2. Given this condition certain operations may now be considered.

Sign Changing or Inversion

Clearly if Z_1 and Z_2 are equal resistors $A_0 = -1$, i.e. the output is the inverse of the input. The circuit may be drawn as in Fig. 6.

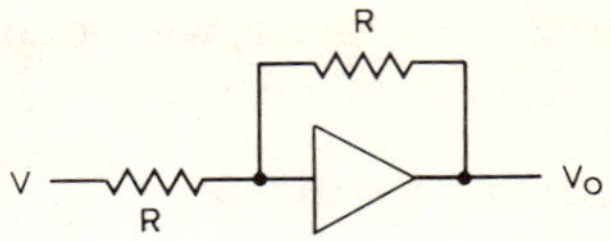

Fig. 6.—*Operational inverter.*

Scale Changing

If Z_1 and Z_2 are unequal resistors, i.e. $Z_2/Z_1 = K$, the input is multiplied by $-K$ at the output; K may be greater or less than 1.

Addition

In Fig. 7, with equal input resistors, I and therefore V_0 will be proportional to the sum of the input voltages V_A, V_B, V_C etc.

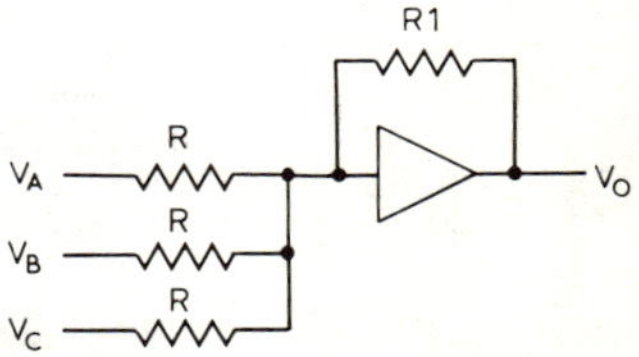

Fig. 7.—*Operational adder.*

Differentiation and Integration

With suitable values of R and C, V_0 in Fig. 8 (a) and (b) will be the differential and integral, respectively, of V.

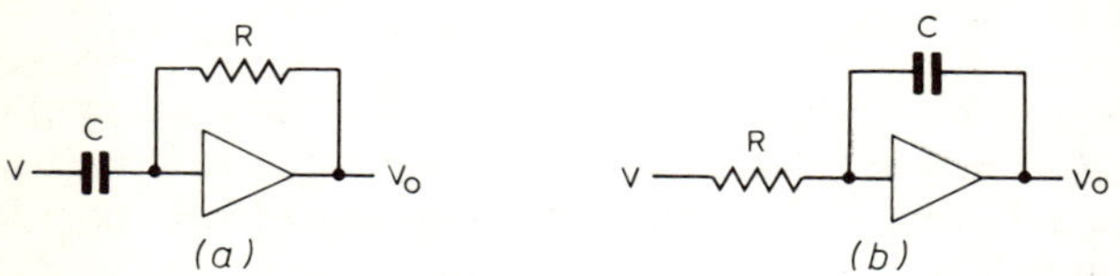

Fig. 8.—(*a*) *Operational differentiator.* (*b*) *Operational integrator.*

Choppers

The technique here is to *chop* (i.e. sample or modulate) the slowly varying input waveform into an a.c. waveform, amplify the a.c. in a conventional amplifier and then demodulate the resultant back to its original form.

The simplest form of chopper is a mechanical switch, but clearly this limits the chopping frequency. If this is too low the chopped waveform may not accurately follow changes in the input waveform. For this reason, and for general convenience, transistor switches are usually used although they introduce an offset voltage.

Fig. 9 illustrates the principle in its basic form, simply showing the chopper as switches S_1 and S_2. When S_1 is closed C_1 charges up to V. When S_1 opens, since the voltage across

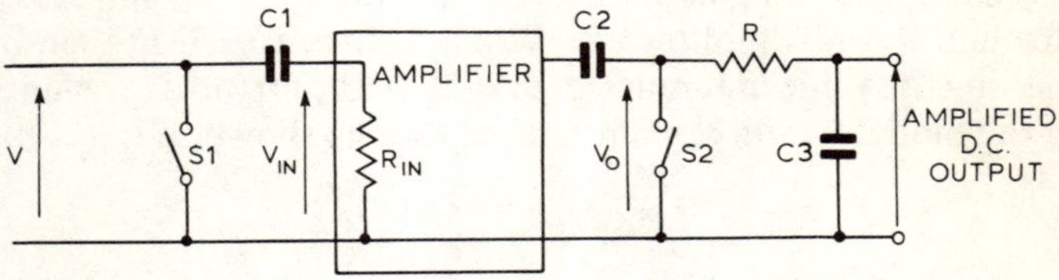

Fig. 9.—*Chopper amplifier.*

C_1 cannot change instantaneously, the voltage across R_{in} rises by V. So, provided the chopping frequency is high, the pulses forming V_{in} follow V. If S_2 is synchronous with S_1, V_0 will be V_{in} amplified and this will be smoothed by R and C_3 to give a d.c. output.

A simple f.e.t. chopper is shown in Fig. 10. The switching waveform amplitude and the value of R_L must be chosen to saturate the f.e.t., and if R_2 is low the offset voltage will be small.

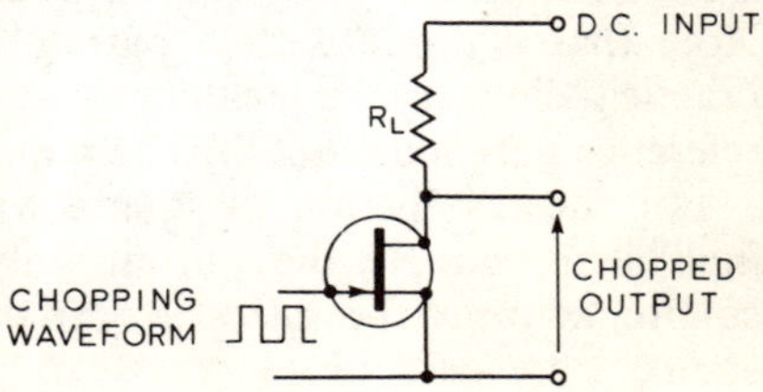

Fig. 10.—*F.E.T. chopper.*

8

PULSE CIRCUITS

Clippers

By slicing the peaks off a sine wave it approximates to a square wave. This operation is accomplished by a clipper, the simplest form of which is shown in Fig. 1 (a). The input waveform V will not bring the diode D into conduction until it exceeds E. As long as D does not conduct no current passes through R and V_0 follows V. When V does exceed E, then by making the value of R much greater than the forward resistance of the diode, V_0 will remain limited to E as shown in Fig. 1 (b).

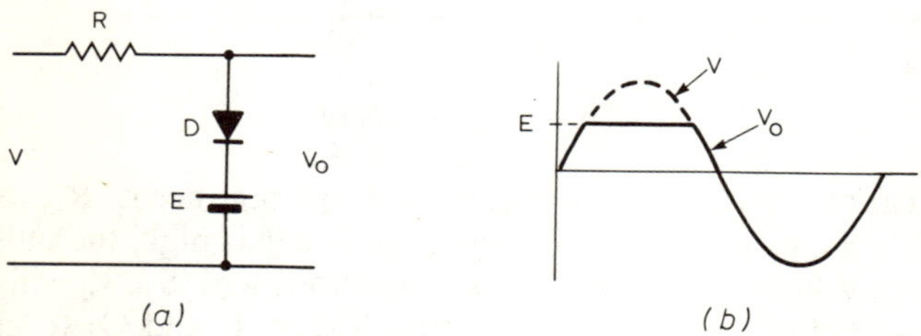

Fig. 1.—(*a*) *Positive peak clipper*. (*b*) *Waveform*.

To clip the negative peaks of V as well a second diode circuit is added as shown in Fig. 2 (a) to give the waveform of Fig. 2 (b). Note that a diode with a particular bias potential can, in practical circuitry, be ideally replaced by a zener diode with a reference potential equal to the required clipping level.

For the production of square waves the diode clipper is inefficient because the output only begins to approach a square wave as the ratio of output to input becomes quite small.

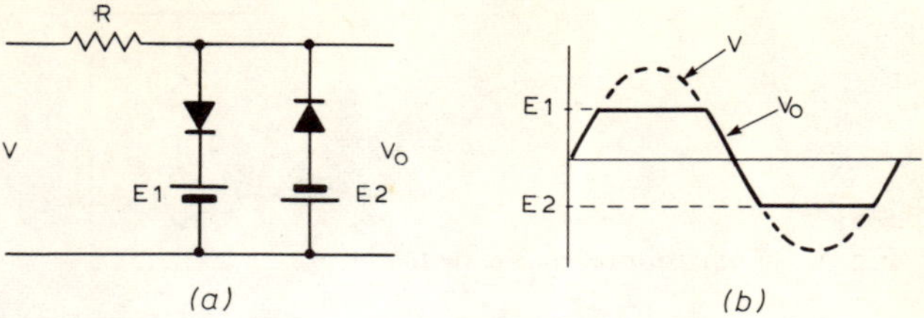

Fig. 2.—(*a*) *Double peak clipper.* (*b*) *Waveform.*

Transistor as a Switch

The basic function of a switch is, as shown in Fig. 3, to connect or disconnect some supply voltage V to a load R.

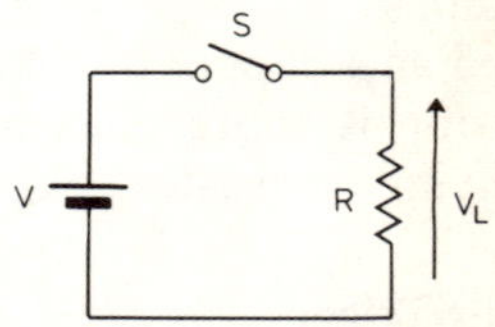

Fig. 3.—*Basic switch circuit.*

The ideal switch passes no leakage current when it is open and is a perfect short circuit when closed. Further, when opened or closed the resultant voltage change should be instantaneous. A switch with these ideal properties will produce the waveform in Fig. 4. When S is open, $V_L = 0$. When S is closed, $V_L = V$. When S changes over, V_L rises or falls instantly.

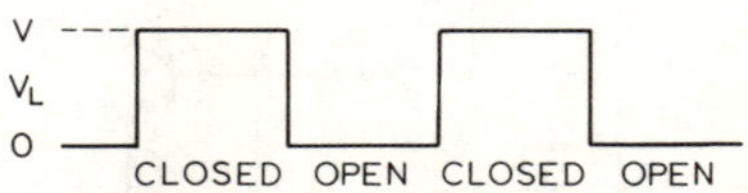

Fig. 4.—*Load waveform for perfect switch.*

As a switch the transistor approaches the ideal. Consider Fig. 5 where the object is to use Q to switch the supply voltage V_{CC} to and from the load resistance R. If Q can be provided

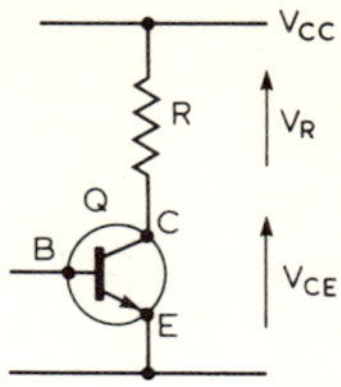

Fig. 5.—*Transistor used as a switch.*

with enough base current to saturate it, V_{CE} approaches 0 V and V_R approaches V_{CC}. Conversely, if the base voltage is reduced to 0 V, or even better made negative, Q only passes leakage current, and so V_{CE} approaches V_{CC} and V_R approaches 0 V. The speed of transition of the output waveform can be made fast by careful choice of transistor and operating conditions.

Used as a clipper by driving the base with a sine wave the transistor is more efficient than the diode because of its amplifying properties.

Schmitt Trigger

A far superior circuit for converting a sine wave into a square wave is the emitter-coupled bistable multivibrator—usually known as the *Schmitt trigger* circuit—shown in Fig. 6.

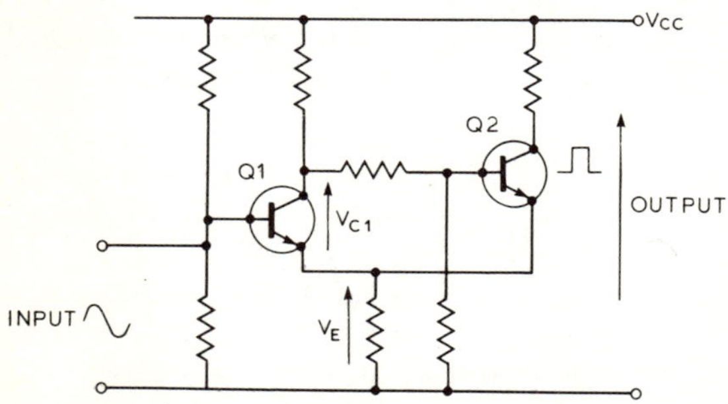

Fig. 6.—*Schmitt trigger circuit.*

With no input the bias resistors are chosen so that Q_1 is off and Q_2 on. When the input signal becomes positive enough to bring Q_1 on, V_{C1} falls and reduces the current in Q_2, which in turn makes V_E fall. This has the effect of making the base emitter voltage of Q_1 more positive and the circuit very quickly changes over to the state Q_1 on, Q_2 off. When the input signal goes negative the reverse transition occurs. The waveforms are shown in Fig. 7.

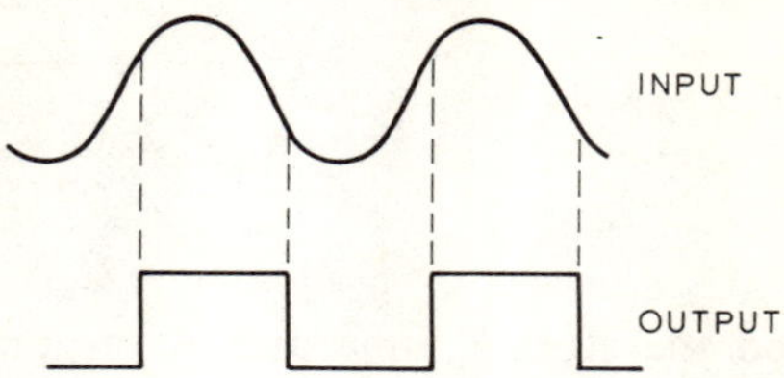

Fig. 7.—*Schmitt trigger waveforms.*

Rise Times of Pulse Circuits

The rise time of a pulse (and here this may include negative-going pulses) is defined as the time it takes for a pulse to rise from 10 to 90 per cent of its maximum value. In such circuits as the Schmitt trigger, where a regenerative switching action is used rather than the clipping technique, the main limitation of the speed of a pulse transistion is the performance of the transistor itself.

Considering an npn transistor where (Fig. 8) waveform (a) is the base voltage waveform used for switching the transistor on and off, then (b) will be the base current waveform and (c) the collector voltage, i.e. the output pulse waveform. The switching delays may be separated into four distinct periods:

T1 is the time taken for holes injected from emitter into base to cross the base region and to start the rise in collector current.

T2 is the rise time for collector current and is limited by the depletion layer capacitance.

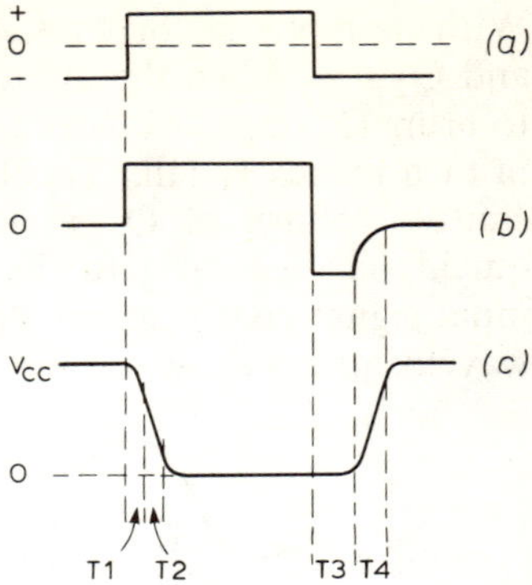

Fig. 8.—*Transistor switch waveforms: (a) base voltage; (b) base current; (c) collector voltage.*

T3 is called the storage time and is responsible for the greatest delay. If more base current is supplied than the minimum necessary to saturate the transistor, excess charge will be stored in the base. When V_{BE} goes negative, collector current will continue to flow until the excess charge has been drained away from the base.

T4 is the fall time corresponding to T_2.

D.C. Restoration

Many waveforms have a d.c. component. For example a transistor switch producing a square wave switching between V and 0 has a d.c. component of $V/2$. If this waveform were passed through an a.c. coupling the d.c. component would be lost and the waveform would switch between $+V/2$ and $-V/2$. In certain circumstances it would be necessary to restore the d.c. component.

Consider a square wave of peak-to-peak value with no d.c. component at the input to the d.c. restorer circuit in Fig. 9. On the first positive transition D conducts, C charges up to V and V_0 is held at 0 V by the diode. As soon as the waveform goes negative D cuts off and, with V_C held constant with the charging path now broken, V_0 falls with the input by V.

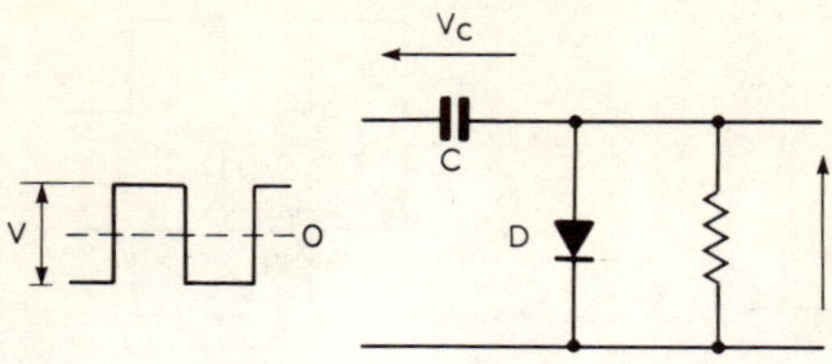

Fig. 9.—*D.C. restorer.*

When the input goes positive V_0 returns to 0. The waveforms are shown in Fig. 10. It can be seen that the output will vary from 0 to $+V$ if D is reversed.

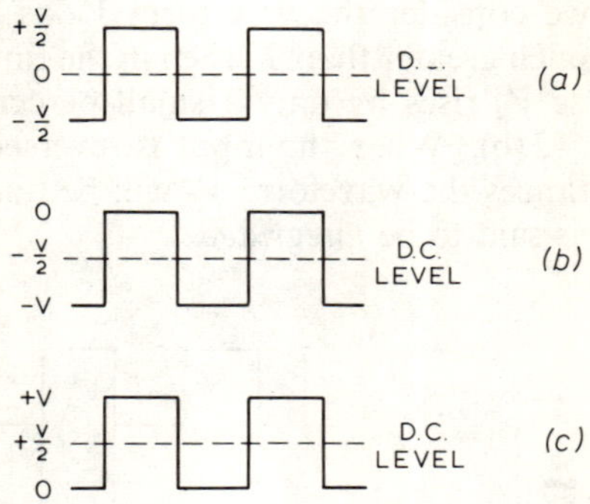

Fig. 10.—*D.C. restorer waveforms:* (*a*) *input with no d.c. component;* (*b*) *output with D as in Fig. 9;* (*c*) *output with D reversed.*

Differentiating and Integrating Circuits

Consider a voltage square wave being fed to the series CR circuit of Fig. 11 (a). At the instant the pulse goes positive V_R increases by V, the amplitude of the pulse as shown in Fig. 11 (b). If CR, the time constant of the circuit, is much less than T (the time of a half cycle of the waveform) then C will quickly be charged up so current will stop flowing and V_R will fall to 0. When the input goes negative the voltage across C cannot change instantaneously so V_R falls by V, but again

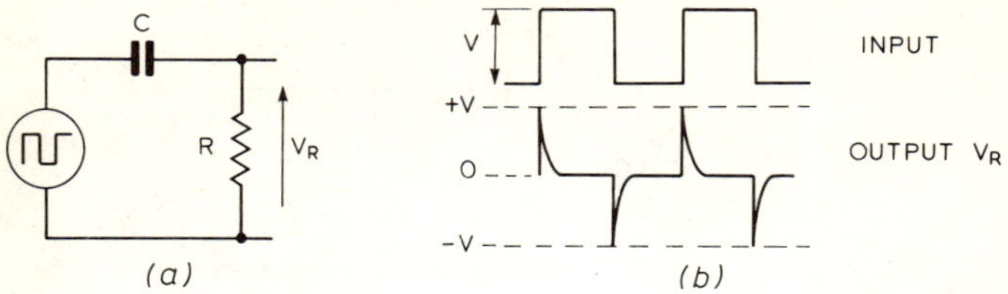

Fig. 11.—(*a*) *CR differentiating circuit.* (*b*) *Differentiating circuit waveforms.*

C soon assumes the value of input voltage and V_R falls to 0 again. The input waveform is said to be *differentiated* at V_R.

Instead, if we consider the waveform V_C as in Fig. 12 (a) (where CR is much greater than T) then in the time duration of a positive pulse V_R rises by only a small percentage of V as shown in Fig. 12 (b). When the input is reversed V_C falls. In these circumstances the waveform V_C will be quasi-triangular and the input is said to be *integrated.*

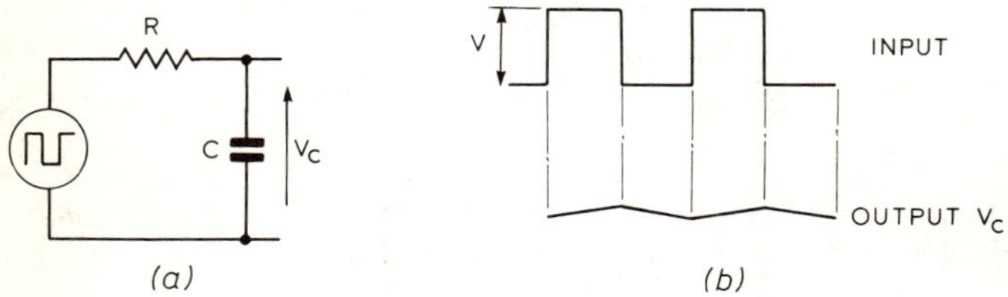

Fig. 12.—(*a*) *CR integrating circuit.* (*b*) *Integrating circuit waveforms.*

The degree to which the differentiation and integration process approach the ideal is determined by the relative values of CR and T. In the case of the differentiator the effective value of resistance used will be reduced by using the operational amplifier technique mentioned in Section 7. The circuit for differentiation is shown in Fig. 13 (a), where the effective value of resistance in series with C will be approximately R/A. For

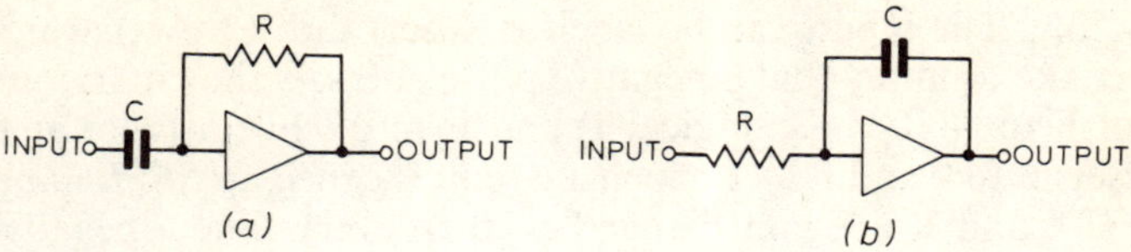

Fig. 13.—(*a*) *Operational differentiator*. (*b*) *Operational integrator*.

integration the circuit of Fig. 13 (b), sometimes called the Miller Integrator, in effect increases the circuit capacitance to $C \times A$ and this gives a more linear waveform than simply using the time constant CR.

Bistable Multivibrator

The collector-coupled bistable multivibrator shown in Fig. 14 is sometimes called the *Eccles Jordan* circuit, the

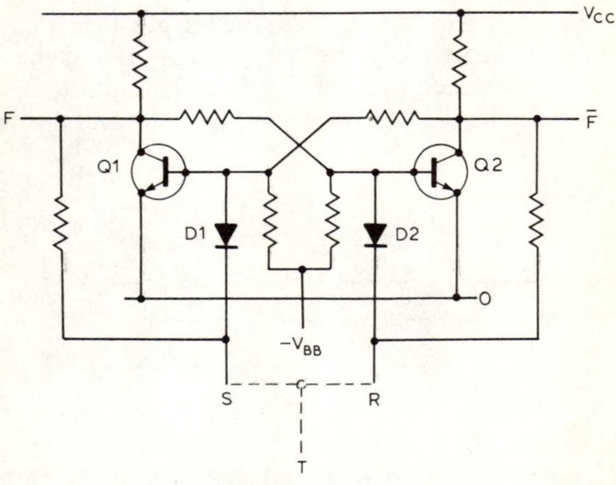

Fig. 14.—*Bistable multivibrator.*

flip-flop (FF) or the *binary*. Suppose it is in the state Q_1 on, Q_2 off. Then a negative pulse onto Q_1 base will switch it off and the circuit will make the transition to the state Q_1 off,

Q_2 on. The inputs can be labelled S (set) and R (reset), which we take to imply that a negative pulse at S sets the binary output F to +10 V (i.e. logical 1) and $\bar{F}$ to 0 while a pulse at R resets F to 0 and $\bar{F}$ to 1. Such a circuit is called an *SR flip-flop*.

If S and R are joined together, then every time a negative pulse is applied to the input, which we now label T, the binary changes over. This may be seen as follows. Suppose the circuit is at $F = 1$, $\bar{F} = 0$ (i.e. Q_1 off, Q_2 on). Then D_1 will be cut off by the collector voltage of Q_1 being held positive, while D_2 has no voltage across it and will conduct for a negative pulse at T. This will cut off Q_2 so the circuit changes over. Now D_1 only will conduct and the next negative pulse will cut off Q_1 and so on. The diodes are called *steering diodes* because they steer the trigger pulse to the transistor where it will be effective.

Binary Division

With an input T as in Fig. 15 (a) the output at F (Fig. 15 (b)) will have one negative edge for every two at T, and the binary is said to have divided by 2. If F now is the input to a second binary its output will be as shown in Fig. 15 (c) and the original

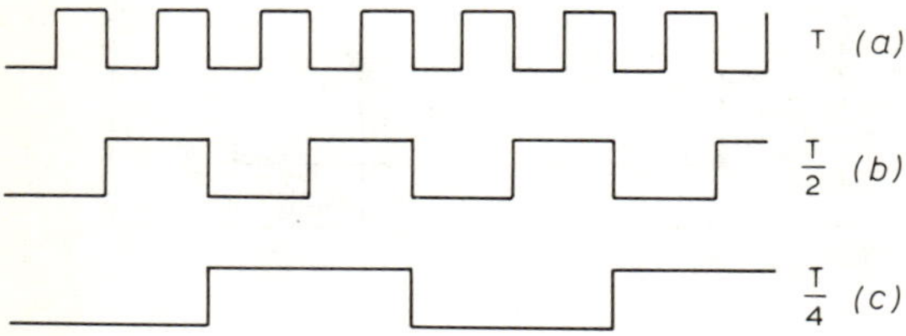

Fig. 15.—*Binary chain waveforms: (a) binary 1 input; (b) binary 1 output ≡ binary 2 input; (c) binary 2 output ≡ binary 3 input.*

input pulse train has been divided by 4. In general, *n* stages divide by 2^n; e.g. with 3 stages, the division ratio would be $2^3 = 8$; 5 stages, $2^5 = 32$, etc.

Non-binary Division

To divide by numbers other than 2^n the division rate of a binary chain can be reduced by feedback, sometimes called *knockback*. For example the three stages of Fig. 16 without

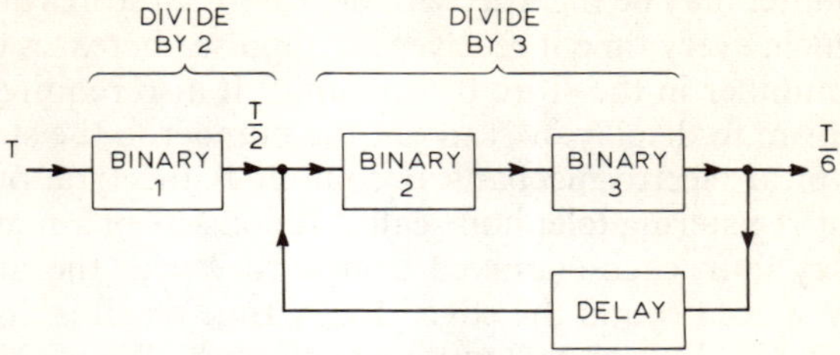

Fig. 16.—*Feedback reducing division rate from 8 to 6.*

feedback divide by 8. A pulse fed back from the output (delayed to avoid confusion between the input and fedback pulses) produces the same effect on the middle binary as two pulses into the chain. It now takes only six input pulses to produce one output pulse, so the circuit divides by 6. As a general rule we can say that if the feedback loop excludes x stages (in Fig. 16 this will be one—the input stage) it reduces the division rate by 2^x. Thus, to divide by 10 we should use four stages as in Fig. 17. Loop A excludes one stage, therefore

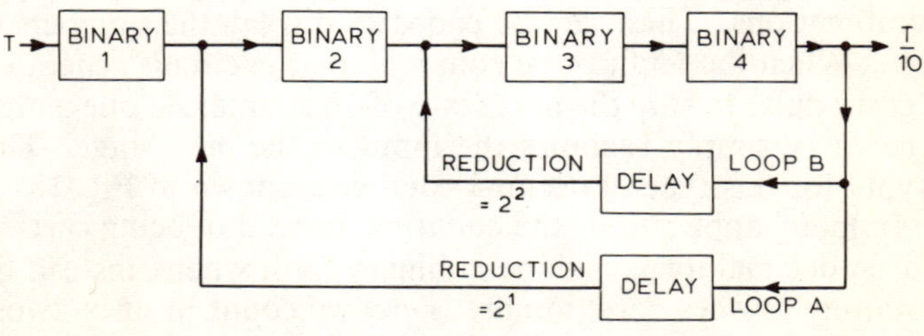

Fig. 17.—*Divide-by-10 circuit.*

the reduction is $2^1 = 2$. Loop B excludes two stages, so the reduction is $2^2 = 4$. Therefore the total reduction is $2+4$ and the division rate is $16-6 = 10$.

Counters

A counter may be regarded as a device which stores numbers and which, every time it receives an impulse, increases the size of the number in the store by one unit. It also requires some mechanism to display or read out the number in the store.

A typical electro-mechanical counter is the type of meter used for registering telephone calls. It consists of a number of dials, say four, each engraved from 0 to 9 with the numbers equally spaced round the edge. Every time a call is made the right hand dial is stepped on by one number. When the meter reads 0009 the right hand dial moves on to 0 and in so doing moves the next dial to 1, i.e. it makes a carry so that the meter reads 0010. The total store capacity is 9999. The dials continuously display the total count. See the end of this section for further details of mechanical counters.

For high-speed computing, purely electrical methods are used for counting. We have seen that binary circuits used in cascade can divide by any number, and for the traditional decimal counting system a series of divide-by-10 circuits could be used, one to represent each integer in the number. In such an application the divider is called a *scale-of-*10 *counter*. Over a series of ten input pulses the output of the four binaries making up a scale-of-10 counter will go through ten ùnique combinations. These can be coded to display the numbers 0 to 9. When the tenth pulse comes along the circuit generates a carry pulse to stop the next scale-of-10 counter on one plane. The carry simply becomes the input to the next stage. The layout for this type of decimal counter is shown in Fig. 18.

In many applications the counting, instead of being carried out in decimal form, is done in binary form where, instead of counting in ones, tens, hundreds etc. we count in ones, twos, fours, eights etc. More details of this system are given in Section 10.

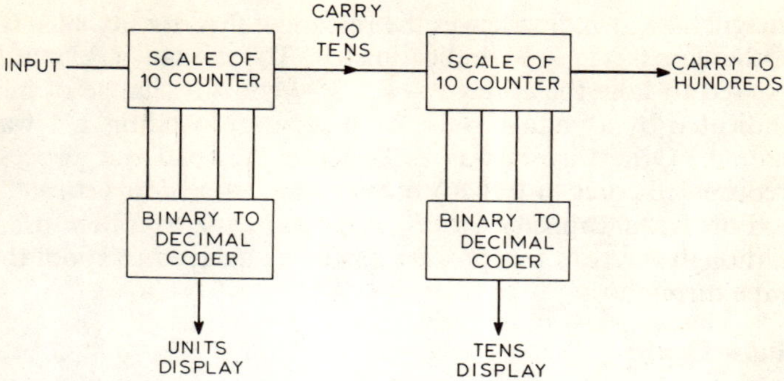

Fig. 18.—*Counter giving a decimal output.*

Stores

In the counter of Fig. 18 the binaries themselves store the numbers. All a binary can do is to change over when it receives a pulse, and in so doing it is said to *store* one *bit* (binary digit). Most other electrical stores also work in this way, i.e. instead of storing ten numbers (0–9) as in the telephone call meter they store only one.

Magnetic Stores

Another element for storing a bit of information is the toroidal magnetic core shown with the type of hysteresis loop required in Fig. 19. When storing a 1 the current in a

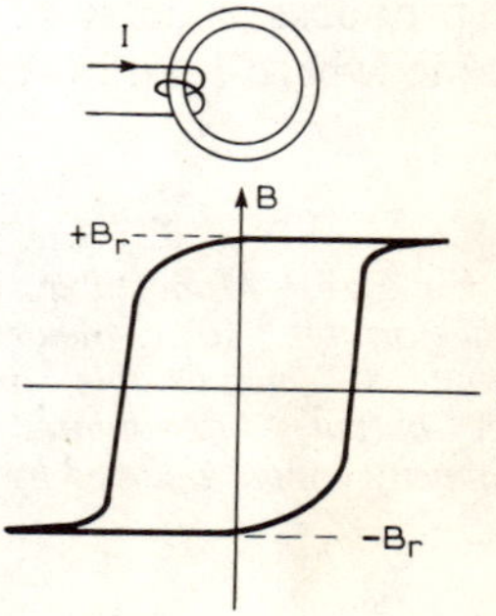

Fig. 19.—*Toroidal magnetic core and hysteresis loop.*

magnetising winding leaves the remanent flux density at $+B_r$. Otherwise it is at $-B_r$, indicating a 0. To read out, a current is passed to take the core to $-B_r$. If there is a change of flux indicated by a voltage pulse on a separate winding, a 1 was stored. Otherwise it was a 0. Since the read out process reduces all cores to 0 the process is said to be *destructive.*

For large capacity stores magnetic tape is often used, although the read out process has to be delayed to spool the tape through.

Pulse Delays

It is often necessary to delay pulses, for example in the feedback loops of divider circuits. One method of delay is simply to use a piece of transmission line. If a specimen is chosen which is comparatively loss-free, it may be considered as an LC ladder network as shown in Fig. 20. The reactive

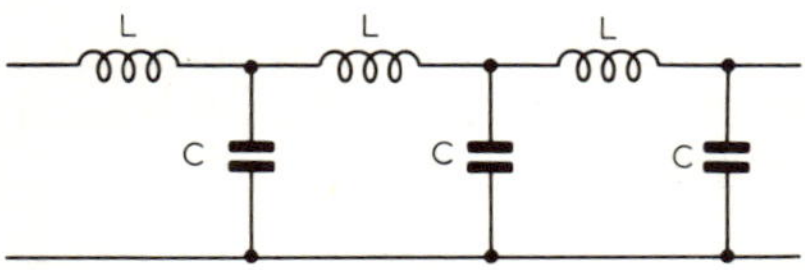

Fig. 20.—*Equivalent circuit of a loss-free line.*

elements retard the phase of signals passing down the line and it produces a delay of $1/\sqrt{LC}$ seconds per metre length of cable, where L and C are the inductance and capacitance per

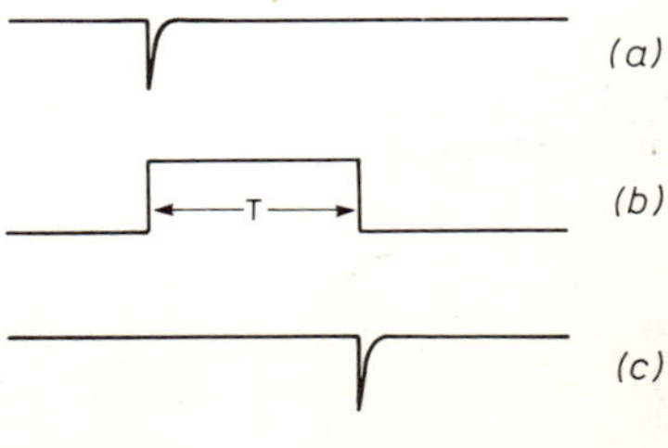

Fig. 21.—*Monostable delay waveforms: (a) monostable input; (b) monostable output; (c) output differentiated and clipped = input delayed by* T.

metre of cable. In practice, delays of more than a few microseconds require a considerable bulk of cable.

An alternative is the monostable multivibrator mentioned in Section 6. Suppose that this, in response to a negative pulse (Fig. 21), produces an output pulse of duration T. By differentiating this pulse and then clipping the positive peak, there remains a negative pulse similar to the input but delayed by T.

Phase Detection and Correction

To establish a locked phase relationship between a reference pulse train A and a generated pulse train B, a circuit such as Fig. 22 (a) can be used. Pulse train B triggers a sweep generator to produce the waveform of Fig. 22 (b). This is sampled at a

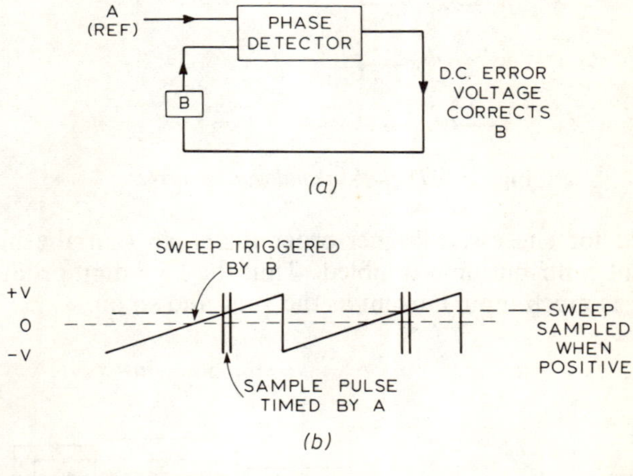

Fig. 22.—(*a*) *Phase detector and correction layout*. (*b*) *Phase detector waveforms*.

time determined by A. The d.c. level out of the sampling circuit is determined by the time at which it is sampled. If it is as shown in Fig. 22 (b) the output will be positive and this voltage can be used to adjust the frequency of B until the required phase relationship is established.

Digital to Analogue Conversion

Where a number is represented digitally, say in binary form, there is sometimes the requirement to convert it to analogue form. For example the binary number 1101 (which is decimal 13, i.e. $1 \times 8 + 1 \times 4 + 0 \times 2 + 1 \times 1$) could be represented by the state of four binaries, but it may be required in analogue form as one quantity, say 13 volts. This may be achieved by the operational adder of Fig. 23 where the input resistor value is

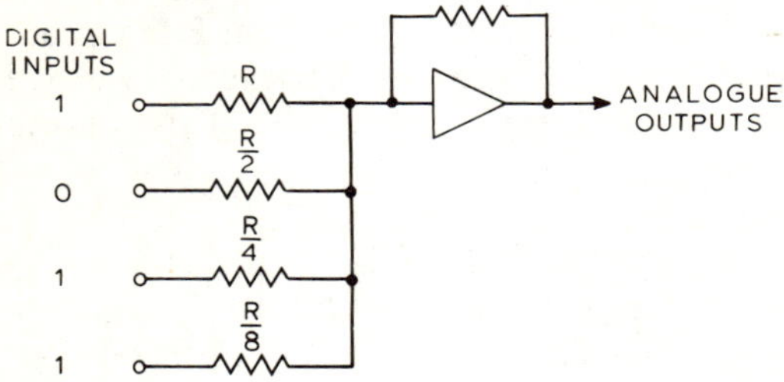

Fig. 23.—*Digital-to-analogue converter.*

halved for successive higher-order digits, so that the input current contribution is doubled. Thus the 1×8 digit produces twice as much input current as the 1×4 and so on.

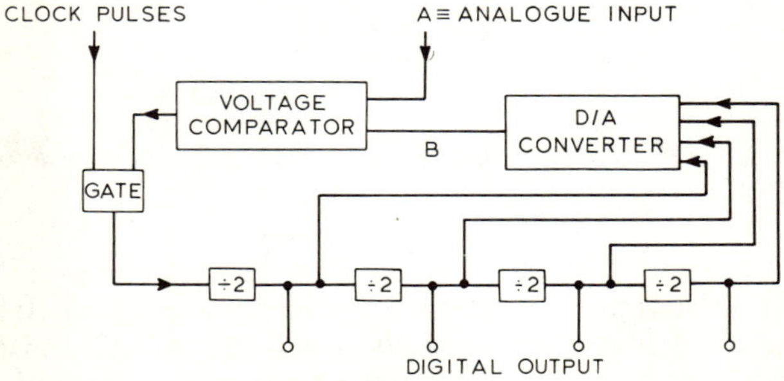

Fig. 24.—*Analogue-to-digital converter.*

Analogue to Digital Conversion

Fig. 24 illustrates analogue to digital conversion. An analogue quantity A is fed into a voltage comparator together with the output of a D/A converter. The comparator only produces an output when the D/A converter reaches the same level as A. When this happens it closes the gate, but until this time clock pulses pass through the gate and are stored by the binary counter. The counter output is converted to an analogue quantity B and when this equals A the gate is closed and the binary counter state represents the analogue signal A.

Pulse Modulation Systems

The principle of time-sharing of communication channels and the development of pulse circuit techniques has led to the increasing use of modulating signals into pulse trains, as opposed to the more traditional methods of amplitude modulation (a.m.) and frequency modulation (f.m.). Four methods of pulse modulation are illustrated in Fig. 25.

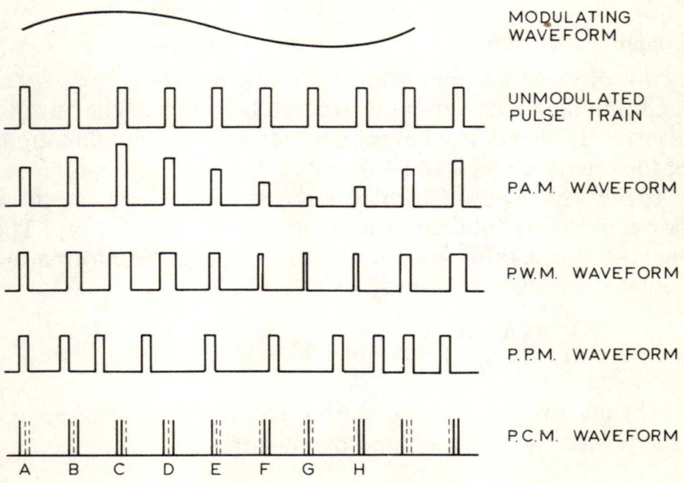

Fig. 25.—*Pulse modulation signals.*

(1) In *pulse amplitude modulation* (p.a.m.), the amplitude of the pulse varies with the instantaneous value of the modulating signal.

(2) In *pulse width modulation* (p.w.m.), sometimes called *pulse duration modulation* (p.d.m.), the width of the pulse varies with the instantaneous value of the modulating signal.

(3) In *pulse position modulation* (p.p.m.), the time at which a pulse occurs varies with the instantaneous value of the modulating signal.

(4) In *pulse code modulation* (p.c.m.), the instantaneous value of the modulating signal is represented not by one pulse but by a train of pulses (in this diagram, three). Each time a pulse train occurs it indicates the value of the waveform as a binary number. In this diagram a 1 is represented by the presence of a pulse (solid line) and a 0 by its absence (dotted line). The points labelled, then, are as follows:

A	B	C	D	E	F	G	H
100	101	110	101	100	011	010	011
4	5	6	5	4	3	2	3

Channel Capacity

It is obviously of importance that the maximum capacity of a communications system is accurately known. The various levels of the signals can be represented by various combinations of the binary digits 0 and 1 for any form of transmission.

If the signal power available is given by S and the noise power is N the total received voltage will be $\sqrt{(S+N)}$. The signal-to-noise ratio will be $\sqrt{[(S+N)/N]}$, and the information content of a channel is given by

$$H = \log_2 \sqrt{\left(\frac{S+N}{N}\right)} \text{ bits} \quad \text{or} \quad H = \log_2 \sqrt{\left(1+\frac{S}{N}\right)} \text{ bits}$$

The maximum number of pulses that can be sent per second, n, is related to the period and bandwidth:

$$n = \frac{1}{t} = 2B \text{ pulses}$$

where t = period and B = bandwidth; the maximum information content of a channel is defined as

$$H = BT \log_2 \left(1 + \frac{S}{N}\right) \text{ bits}$$

This is the Hartley–Shannon law of information. From it the signal-to-noise ratio of a channel can be evaluated for particular transmission systems (e.g. p.c.m., a.m., etc.), or the effect on signal-to-noise ratio caused by restriction of bandwidth can be seen.

Mechanical Counters

Fig. 26 is a simplified diagram of an electro-mechanical counting device. An input signal in the form of a current pulse will operate the relay which will draw in the ratchet and cause the first—or "units"—number drum to rotate a few degrees and display a number equal to an increase of one over the number it was already showing. In this way input pulses are counted until the drum has rotated 360 degrees, which is usually equal to ten, and a small protrusion on the first drum

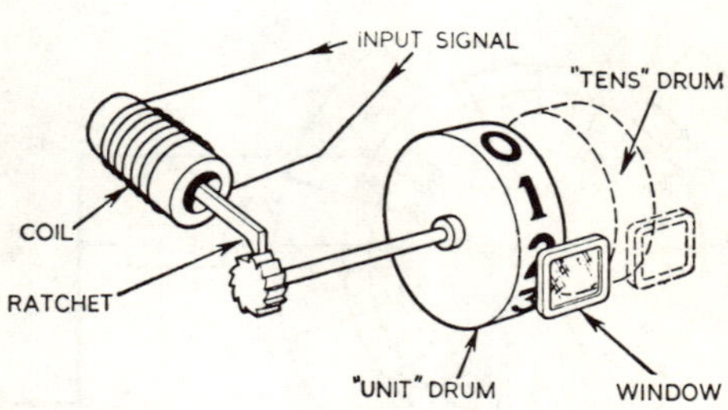

Fig. 26.—*Simple electro-mechanical counter.*

then makes the second—or "tens"—drum move one place. Thus one place on the second drum is equal to ten places on the first. In this way large numbers can be quickly counted and displayed.

These devices find many current applications. One of the latest can be preset to give an electrical output at any number, for example 82,530. Such devices are very useful in automatic processes.

Dekatrons

The Dekatron is a cold-cathode glow-transfer tube. Its operation can be explained by reference to Fig. 27. The central anode is positive with respect to the ten cathodes around its periphery. Between each cathode is a pair of electrodes called guide pins which are normally held at some potential slightly above that of the cathodes. All the "A" pins are connected together and all the "B" pins are similarly connected. The "B" pins are connected to the same input source as the "A" ones except that in their case a very small delay is introduced. When a high potential is applied to the tube, one of the anode cathode gaps is ionised (the glow is visible through the glass envelope). Let us assume that the ionisation is between the anode and the first cathode, as shown. This ionisation will automatically lower the anode potential so that the other

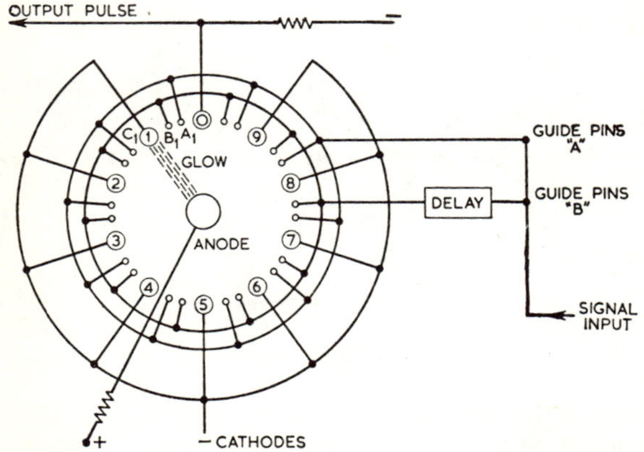

Fig. 27.—*The Dekatron counting tube. One output pulse is given for every ten input pulses.*

cathodes are excluded from the action. The glow has no incentive to move from the cathode it is "stuck" on especially as the guide pins are slightly positive with respect to the cathodes. Consider what will happen when a large negative signal is applied to the guide pins: first the "A" pins become very negative and offer a better potential difference to the anode than does the first cathode, hence the glow jumps to guide pin "A1". A microsecond or so later, the negative signal makes the guide pins "B" negative; the glow will then move to pin "B1". If next the negative signal ceases, the guide pins will return to their normal slightly positive state. The glow cannot return for the same reason that it could not go forwards before the arrival of the negative signal pulse, and therefore has no alternative but to move to the second cathode.

Thus by applying successive negative pulses to the guides the glow discharge passes from cathode to cathode in a clockwise or additive direction. By inserting the delay in the "A" pin line instead of the "B" pin line, the discharge will move in an anti-clockwise or subtractive direction. Thus a series of pulses are counted and displayed by the Dekatron. When the ten cathodes have in turn been ignited by ten signal pulses, it can be arranged that the tenth or output cathode will produce a signal which can be used to drive a further Dekatron decade counter. In Fig. 27 a load resistor is shown connected to the output cathode.

A series of tubes arranged in this way enables us to count units, tens, hundreds, thousands and so on and to display the final result. Dekatrons will be found in many applications from medical electronics to nuclear engineering. To some extent they can be used as a "store" in computers, although more efficient methods of storing numbers do exist.

Vacuum Counting Tubes

Vacuum counting tubes are available which retain the direct indication advantages of the cold-cathode gas-filled decade counters, but are capable of much higher counting rates, limited only by the associated circuitry.

An example of this type of tube is the Mullard E1T ribbon counting tube, the basis of which is a ribbon-shaped electron beam which can be deflected into ten well-defined positions by the input signals. In any of these positions the beam passes through one of ten apertures in a cylindrical anode and impinges on a fluorescent screen causing a spot of light to appear opposite the appropriate figure (0 to 9) marked on the tube envelope. As the last position is passed by the beam a signal can be generated to reset the tube to "0" and simultaneously apply a counting pulse to the next tube in the chain.

This tube is incorporated in a series of units manufactured by Mullard Equipment Ltd., for counting applications where it is not required to produce command signals for control purposes.

These units are normally capable of counting speeds up to 30,000 per second, and special add-on units are available to increase the speed to 100,000 per second. The E1T tubes which give a visual indication of the count are driven by monostable multivibrator (see Section 6) circuits.

Apart from the high-speed decades, a number of other plug-on ancillary units are available. These include a pulse-former for use when the input transducer does not provide a pulse of the shape required for operating the counter; a gating unit, which can also be used as a pulse pre-amplifier to increase the sensitivity of the counter; a timing unit incorporating a crystal-controlled oscillator for the precise measurement of time intervals; and an electro-mechanical counting unit, which, for low-speed counting, can be used in place of the last decade unit in a chain to increase the counting capacity by a factor of 1,000.

The Trochotron

For speeds beyond the capacity of the cold-cathode counting tubes trochotron beam switching tubes have been developed. These tubes, which will operate at speeds up to 1 MHz, are vacuum devices making use of crossed electric and magnetic fields to form a beam of electrons between an indirectly heated cathode and any of ten groups of three electrodes mounted around the cathode, Fig. 28. The inter-electrode potentials within the tube provide the electric field, and the magnetic field is due to a cylindrical permanent magnet fitted externally around the glass envelope.

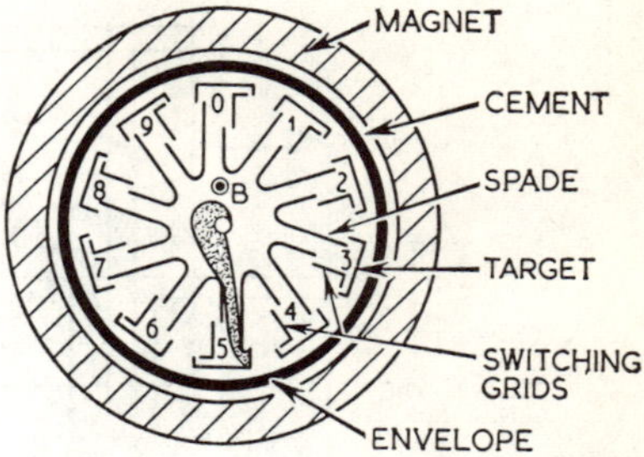

Fig. 28.—*The principle of operation of the trochotron: the electron beam is shown in position* 5.

Each group of electrodes consists of a spade, which forms and locks the beam in position; a target, which makes the beam available as a constant current output; and a grid for switching the beam from one spade to the next. When power is first applied, all spades will be equally positive with respect to the cathode and, due to the action of the magnetic field in preventing electrons reaching the electrode-groups, the tube will be in a "cut-off" condition, with no beam formed. However, if the potential of any spade is reduced, by means of a high-speed pulse or a d.c. voltage, the beam will form on the electrode group associated with that spade, and an output will appear on the corresponding target.

Once formed, the beam is held in this position by a combination of the spade series resistance and spade current until it is stepped to the next position by lowering the voltage on the associated switching grid.

Since only the grid in its immediate vicinity will affect the beam, the grids are connected internally in two groups: the odd-numbered in one, the even-numbered in the other. This makes it possible to use a d.c. input for switching and still obtain a single position stepping, thus avoiding the necessity for a pulse of critical width. The switching grids can conveniently be driven by a bi-stable circuit or push–pull sine-wave drive. A simple drive circuit (bi-stable circuit not included) is shown in Fig. 29.

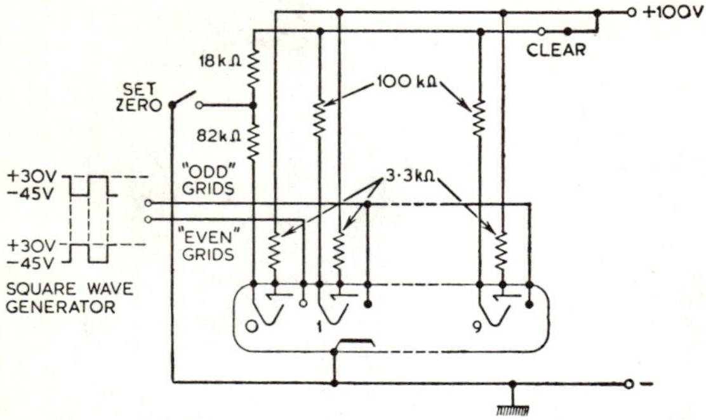

Fig. 29.—*Simple drive circuit for a trochotron. The pulses to be counted first trigger a bi-stable gate (not shown here) which provides the square-wave pulses for the two sets of grids.*

The trochotron provides an output suitable for driving a cold-cathode decimal-indicator tube and compatible with the input requirements of direct read-out digital indicators.

The Ring Counting Circuit

Ring counting circuits operate on the principle that only one of a series of valves of the ring is conducting at any one time, a typical example using cold-cathode tubes in such a way that the glow of the tube in operation gives a direct visual indication of the count. A suitable tube is the Mullard type Z700U, which has been used in a series of units which are capable of counting at speeds up to 120,000 counts per minute. A positive pulse to the trigger electrode initiates a large direct current flow between anode and cathode through the intermediary of a gas discharge. In the ring circuit the trigger electrode of each tube is connected through a bias resistor to the cathode of the previous tube, so that each trigger electrode assumes the cathode potential of the preceding tube, and also via a capacitor to the pulse input line. In this way the circuit functions so that each input pulse will trigger the tube following the last one to conduct.

LED Displays

See Section 3.

9

LOGIC AND LOGIC CIRCUITS

A switching circuit such as a binary can only have two states. These are arbitrarily designated 0 and 1 and can represent either the binary numbers 0 and 1, as in a counter, or the state of some physical process where 0 and 1 could represent "on" and "off", or "true" and "false". A variable in traditional algebra may be used to represent an infinite number of values, but since the inputs and outputs of switching circuits can only have two values it has been found convenient to use *Boolean algebra* for the analysis and synthesis of switching circuits in a logical manner.

BOOLEAN ALGEBRA

Boolean algebra has three basic propositions:

(1) *Negation* or *complementation.* If we have a variable A its complement is called NOT A, usually written as $\overline{A}$ or A′, and the variable is said to be *primed.* Examining all possibilities (i.e. two) it simply means that if A = 0, $\overline{A}$ = 1 and if A = 1, $\overline{A}$ = 0. This may be expressed in a truth table (Table 1).

Table 1.—Truth Table for NOT Function

A	$\overline{A}$
0	1
1	0

(2) *OR function* or *addition.* A function F of two (or more) variables defined as F = A OR B, written A + B, means that if either A or B (or both) is 1 then F = 1. Otherwise F = 0. The truth table (Table 2) summarises this. (We must remember that the word *addition* is used in a definition as part of an

Table 2.—Truth Table for OR Function

A	B	A + B
0	0	0
0	1	1
1	0	1
1	1	1

algebraic system entirely different to traditional algebra and to compare addition in the two systems leads only to confusion.)

(3) *AND function* or *multiplication.* A function F of two (or more) variables defined as F = A AND B, written A . B or just AB, means that if both A and B are 1 then F = 1. Otherwise F = 0. The truth table is given in Table 3.

Table 3.—Truth Table for AND Function

A	B	AB
0	0	0
0	1	0
1	0	0
1	1	1

The expressions A + B and AB are sometimes written $A \cap B$ and $A \cup B$ respectively. These symbols are known as *caps* and *cups* and are taught using these terms quite early in school nowadays.

Following from the three basic propositions are a number of rules of arithmetic and algebra which are useful in simplifying logic functions.

Arithmetical Rules

(1) $1+1 = 1;$ $\quad 1.1 = 1$
(2) $1+0 = 0+1 = 1;$ $\quad 1.0 = 0.1 = 0$
(3) $0+0 = 0;$ $\quad 0.0 = 0$

Algebraic Rules

(1) $A+1 = 1; \quad A.1 = A$

(2) $A+A = A; \quad A.A = A$

(3) $A+0 = A; \quad A.0 = 0$

(4) $A+\overline{A} = 1; \quad A.\overline{A} = 0$

(5) Commutative law

$$A+B = B+A; \quad A.B = B.A$$

(6) Associative law

$$A+B+C = A+(B+C) = (A+B)+C$$
$$ABC = (AB)C = A(BC)$$

These first six rules are self explanatory.

(7) $A+AB = A(1+B)$
$= A$ using rule (1)

(8) $A(A+B) = AA+AB = A+AB$
$= A$ using rule (7)

(9) $A(\overline{A}+B) = A\overline{A}+AB = AB$ using rule (4)

(10) $A+\overline{A}B = A+AB+\overline{A}B$ introducing a redundant term by rule (7)
$= A+B(A+\overline{A}) = A+B$

(11) $(A+B)(A+C) = AA+AB+AC+BC$
$= A(1+B+C)+BC = A+BC$ using rule (1)

(12) $(A+B)(A+\overline{B}) = AA+AB+A\overline{B}+B\overline{B}$
$= A(1+B+\overline{B}) = A$

De Morgan's Theorem

(13) $\overline{A+B} = \overline{A}.\overline{B}$ and

(14) $\overline{A.B} = \overline{A}+B$

This theorem has a special significance apart from its role in simplification of Boolean expression because, for reasons which will emerge in this chapter, it is convenient to manufacture logical elements which implement these functions directly.

The following is an example of the use of rules to simplify an expression:

$$[(A+B')C'+(A'B)C']'$$

$$= [(A+B')C'+(A+B')C'] \text{ using de Morgan}$$

$$= [(A+B')C']' \text{ removing redundant term}$$

$$= [(A'B)'C']' = A'B+C \text{ using de Morgan twice.}$$

LOGIC GATES

In recent years the development of gates for the implementation of logic functions has undergone great progress. It has given rise to abbreviations such as D.T.L., T.T.L. and so on which have become firmly established in the electronic engineer's language. The main parameters which are the designer's concern are:

(1) Speed of response, which is expressed in terms either of propagation delay or of highest operating frequency.

(2) Fan-in and fan-out, i.e. the numbers of identical inputs and outputs a circuit can accommodate without serious deterioration of its performance.

(3) Noise immunity, i.e. the degree to which a circuit avoids giving incorrect output in the presence of noise on supply lines and so on.

Apart from these questions of performance there are such matters as cost, ease of fabrication into i.c. form and power dissipation.

Logic Levels

In determining the function which a circuit implements we must define the logic levels and what they represent. For example, suppose the voltages to be used are +10 V and 0 V. One of these must be defined as 0 and the other as 1. Positive logic is defined as that in which the more positive voltage is 1, so in this case 0 V = 0, +10 V = 1. In this chapter we shall use this logic unless otherwise specified.

Diode Logic (D.L.)

For the circuit of Fig. 1 (a), F = A + B (OR function). This may be seen as follows. If A (or B) is at +10 V (1) with respect to earth its diode will conduct and, provided R is much greater than the forward resistance of the diode, F = 1. F will only be 0 if both A and B are 0 (earthed). The circuit is said to be a *positive OR gate.*

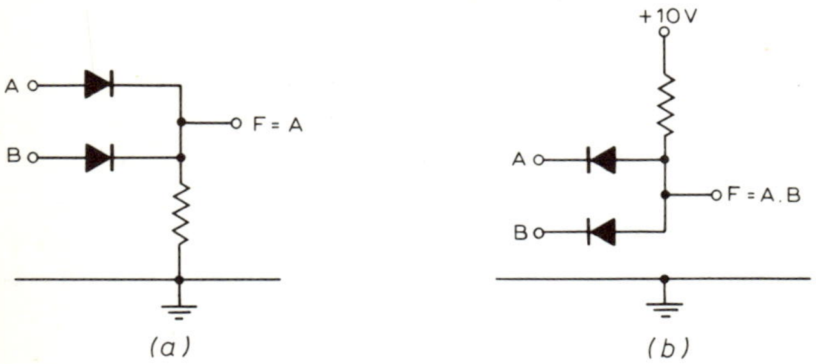

Fig. 1.—(*a*) *Positive OR gate* (*D.L.*). (*b*) *Positive AND gate* (*D.L.*).

The circuit of Fig. 1 (b) is a *positive AND gate.* If both A and B are 1 there will be no current flow around the circuit and F = 1. If one (or both) of the inputs is made 0, the diode(s) will conduct and F will become 0. Therefore F is 1 only when A AND B are 1.

The gate resistors of diode logic circuits should be made large otherwise the input circuits which drive it have to supply a lot of current. On the other hand, if the gate feeds circuits with stray capacitance, the output transition will be slow when the diodes are cut off, i.e. 1 to 0 for OR and 0 to 1 for AND. These factors usually limit the fan out to one.

Logic Conversion

We must now consider what happens if negative logic is used i.e. +10 V = 0, 0 V = 1. All we have to do is draw up the truth table for the gate in positive logic and then draw it again with all 0s and 1s interchanged. The example in Table 4

Table 4.—Interchange of 0s and 1s makes a positive OR gate a negative AND

OR function			*AND function*		
A	B	F	A	B	F
0	0	0	1	1	1
0	1	1	1	0	0
1	0	1	0	1	0
1	1	1	0	0	0

shows that the positive OR gate becomes a negative AND. By a similar process positive AND can be shown to be negative OR.

NOT Circuit

All that is required for a NOT circuit is an inverter such as the common emitter amplifier in Fig. 2. With A = 0 the bias

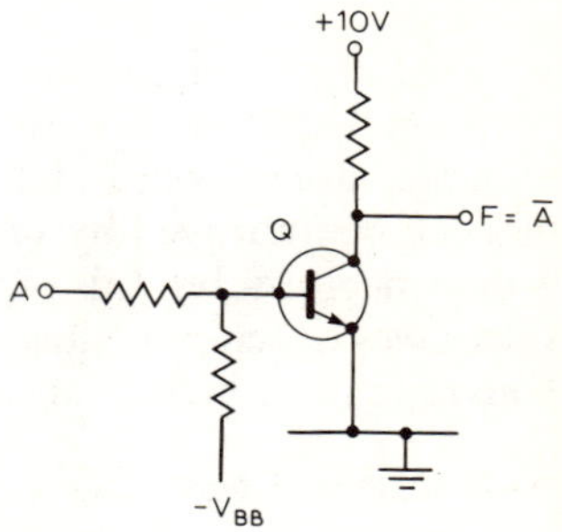

Fig. 2.—*NOT circuit.*

chain holds Q off, so F = 1. With A = 1 the circuit components are chosen to ensure that Q saturates, in which case F = 0. These conditions define F = $\overline{A}$.

Resistor–Transistor Logic (R.T.L.)

This form of logic, shown in Fig. 3, is simply an extended use of the NOT circuit. Its logic function may be determined as follows. If any input is 1, Q will be saturated and F = 0. Only if all inputs are 0 will Q be off and F = 1. The gate function is therefore $F = \overline{A+B+C}$, i.e. NOT (A or B or C), usually described as the *NOR* (NOT OR) *function.* If all inputs except one are earthed (0) then the gate becomes a NOT circuit.

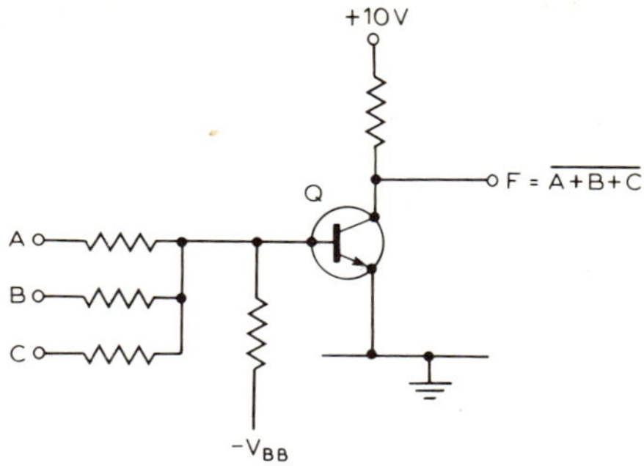

Fig. 3.—*R.T.L. gate.*

The speed of response is slow because Q must be saturated with only one input at 1. If all inputs are 1 the excess charge stored in the base is large and it takes a long time to turn it off. Matters can be improved by shunting the input resistors with speed-up capacitors, but this reduces the noise immunity. The amplifier gives the circuit a lower output impedance than D.L. and consequently a higher fan-out.

Diode–Transistor Logic (D.T.L.)

This type of logic circuit, shown in Fig. 4, has a higher operating speed than R.T.L. If all inputs are 1, point X is positive and Q will be on, making F = 0. With any input at 0, point X falls and the bias chain is designed to switch off Q. Thus $F = \overline{A \cdot B \cdot C}$, usually called the *NAND* (NOT AND) *function.* The speed is faster than R.T.L. because, for the only condition in which Q is on, the circuit can be designed to prevent over saturation of the transistor. In its i.c. form, R is replaced by a pair of diodes (shown dotted in Fig. 4) so that with X at about 0 V, the base of Q is appreciably negative ensuring good noise immunity. To use D.T.L. as a NOT circuit all spare inputs must be held at 1.

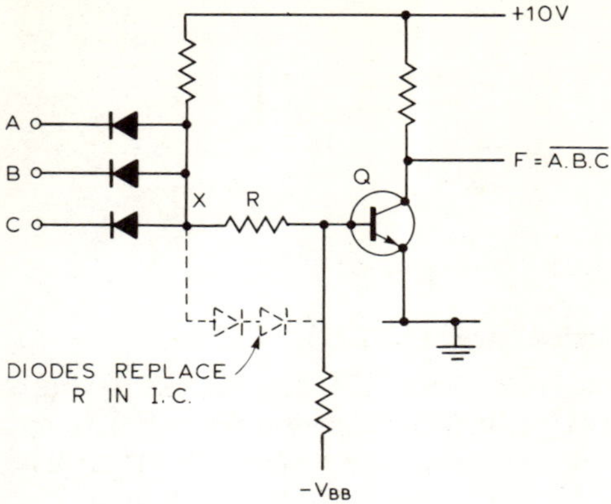

Fig. 4.—*D.T.L. gate.*

Negative NANDS and NORS

Applying similar reasoning to that used in changing to negative logic for the D.L. circuits, it can be seen that R.T.L. is a negative NAND and D.T.L. a negative NOR.

Transistor–Transistor Logic (T.T.L.)

T.T.L. was the first type of logic circuit designed specifically for manufacture in integrated form. Referring to basic T.T.L. circuit of Fig. 5, the multiple-emitter transistor Q1 stands out as a novel feature. In the i.c. form of D.T.L. the path between input and transistor base consists of a pair of diodes back to back. In T.T.L. each emitter to base to collector path leading to Q2 base is equivalent, so it follows that the logic function is the same, i.e. positive NAND. The limiting factor on D.T.L. speed is the time taken to turn the transistor off, and the removal of output transistor base current by the multi-emitter transistor in T.T.L. is more efficient. T.T.L. is regarded as the ultimate in speed for saturated-transistor logic.

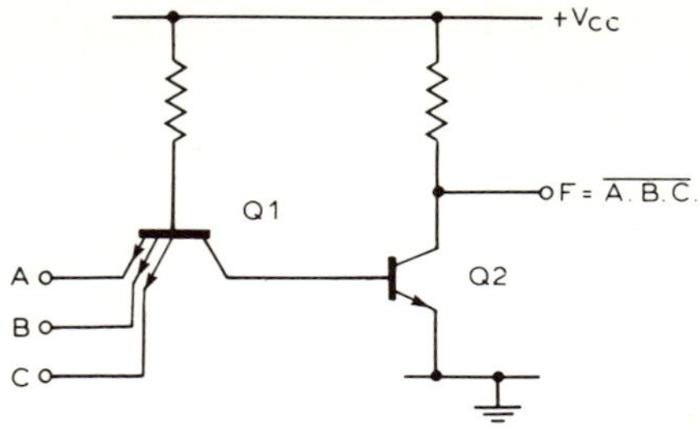

Fig. 5.—*T.T.L. gate.*

Emitter-coupled Logic (E.C.L.)

If we were to reckon a T.T.L. circuit as having a frequency limit of 20 MHz then the non-saturated E.C.L. circuit shown in Fig. 6 would comfortably manage 50 MHz. The difference

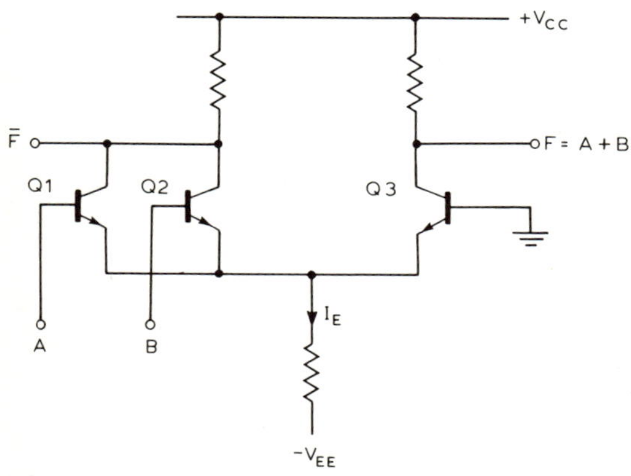

Fig. 6.—*E.C.L. gate.*

in logic levels is quite small, say +0·5 V (1) and −0·5 V (0). With all inputs 0, Q3 is on but not saturated. If any input becomes 1 the emitter voltage rises and I_E is diverted from Q3 to the on transistor(s). This raises F from 0 to 1 so F = A + B.

The other collector will be the complementary function $\overline{F} = \overline{A+B}$.

The circuit only operates satisfactorily with high current values so the power dissipation is large. Fan-in is increased by simply adding extra input transistors up to about 25. Fan-out can also be made high by first feeding the points F and $\overline{F}$ into emitter followers.

Gate Symbols

British Standard symbols for the various gates are shown in Fig. 7.

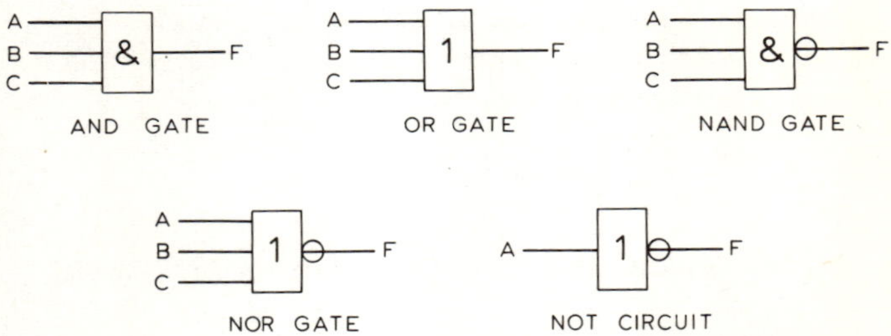

Fig. 7.—*Some British Standard logic symbols.*

Realisation of Basic Logic Functions using NOR or NAND Circuits

The fact that a gate such as R.T.L. implements the NOR function does not restrict the logical functions it can be used to realise. It has been shown that an R.T.L. circuit using only one input is a NOT circuit. Then two NOR circuits connected as shown in Fig. 8 produce the OR function.

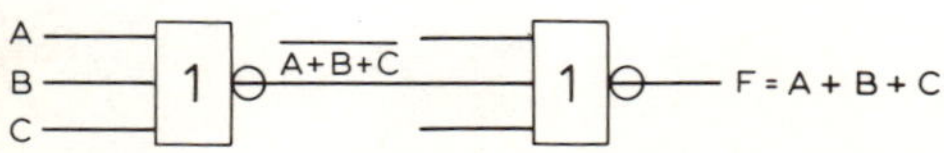

Fig. 8.—*Implementation of OR function using NOR circuits.*

If NOR circuits are connected as shown in Fig. 9 then by using de Morgan's theorem it can be seen that the AND function is implemented, i.e.

$$F = \overline{\overline{A}+\overline{B}+\overline{C}}$$
$$= ABC$$

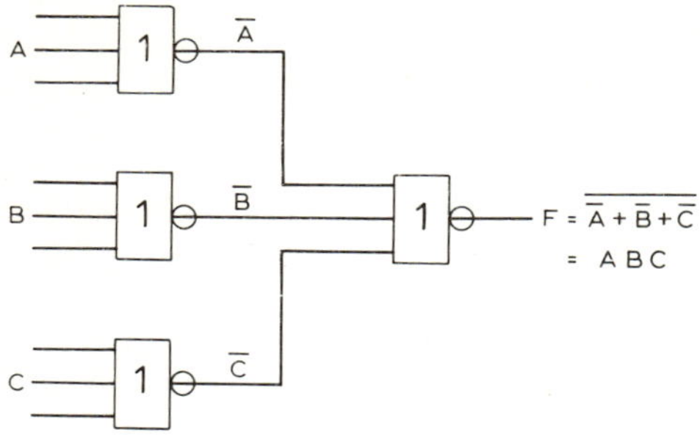

Fig. 9.—*Implementation of AND function using NOR circuits.*

In a similar way NAND circuits can also be used for any of the basic logic functions.

10

ELECTRONIC COMPUTERS

The complexity of calculations associated with modern industry, commerce and research has in many cases increased to a point where solution by normal manual techniques is extremely tedious, difficult or impractical. Electronic computers are being used increasingly in such problems.

Electronic computers can be conveniently divided into two main classes: (1) *digital* and (2) *analogue*. The basic distinction between these classes is the way in which numbers are represented for purposes of calculation. In the digital computer input and output quantities are counted, these quantities being represented by numbers, usually in the binary scale. In analogue computers, numbers are represented by some physical quantity such as length, angular rotation or voltage which can be set for the input of data and measured for output.

Analogue computers are often used for problems concerning quantities which vary with time, and are designed with specific problems in mind; whereas in digital computing, calculations must be broken down into a series of simple steps to make them suitable for the computer, a process called *programming*.

Digital computers are capable of very high accuracy, and this can be increased by enlarging the machines so that they can deal with a greater number of decimal places.

Analogue computers are in comparison low precision devices. However, for the majority of the applications for which they are used, errors of about 1–5 per cent are of little consequence. Analogue systems provide a continuous and simultaneous record of all variables.

DIGITAL COMPUTERS

An automatic digital computer is a machine that will do arithmetic at high speed under the control of *orders* or *instructions* given to it by its operator. The computer is designed to obey a specific number of orders (usually between 30 and 100). To solve any problem the solution must first be written as a series of these orders. Any problem whose solution can be expressed as a series of such simple orders that a computer can obey can be solved on that computer. The reduction of the problem to a series of orders is known as programming, and the actual list of orders is the programme. High computing speeds are attainable because, once the computer has been given the data and programme for a particular problem, it can then compute the answer without any further action by the operator.

Certain features must be incorporated in a computer to enable advantage to be taken of the high speed of the arithmetical unit. First, the computer must be able to obtain its orders and numbers at a speed comparable with that of the arithmetical unit by the provision of a *store* (or "memory") which can hold both orders and numbers until they are needed by the computer, and then extract them at high speed. The store must also be able to receive and hold intermediate computed results.

Secondly, the programme must be so written that it is not necessary to give a separate order for every operation performed, otherwise the speed of the computer would be the speed it could take in information and not the much greater speed of the arithmetical unit. This is done by using repetitive mathematical processes by steps in the calculation; a small group of orders may thus be obeyed many times before computation proceeds to the next step.

The computer itself must be able to determine when it should proceed on to the next step, i.e., it must be able to alter the sequence of orders in accordance with the progress of the calculation. The computer examines the sign of the answer to an arithmetical operation, and enters one of two different storage locations for its next order depending on what sign was found. This power of decision is a very important feature of all automatic digital computers.

In general a computer can be represented by the simple block diagram of Fig. 1. It shows that the computer can be divided into five basic parts: an arithmetical unit, a store, an input device to enable information to be fed into the computer,

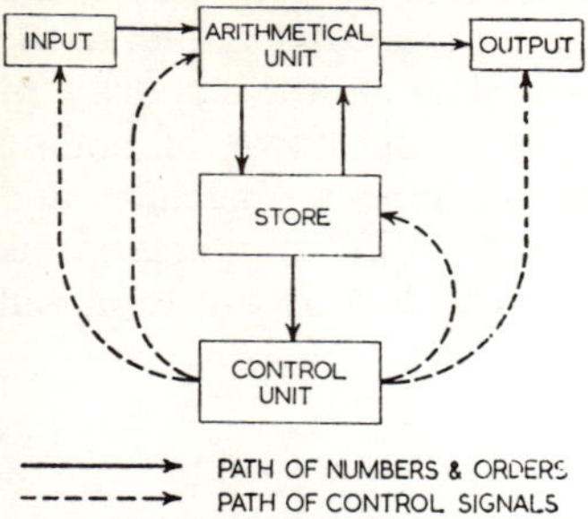

Fig. 1.—*Block diagram showing the basic sections of a digital computer.*

an output device to enable the computer to give out its results, and a control unit to organise the working of the whole machine.

Orders are taken from the store into the control unit which causes the machine to do the operation required. Numbers pass between the arithmetical unit and the store as required. The input device normally feeds into the arithmetical unit, as information fed in must usually be processed in some way before being stored. For a similar reason the output device takes information from the arithmetical unit.

Representing Numbers and Instructions

Digital computers are often classified according to the way that numbers and orders are represented within the machine. Some machines use the decimal code of numbers in which each digit has ten values, 0 to 9. Because of the difficulty of building electronic circuits and storage systems with ten possible states, the vast majority of machines use the *binary* code.

Binary Notation

The majority of everyday calculations are carried out in decimal notation, using figures 0 to 9. With electronic equipment employing relays, transistors, or other on–off or two-state devices, the binary notation, which has only two figures—0 and 1—has the advantage that all numbers can be represented in the form of "on" or "off" states.

The development of pulse techniques employing pulses at a repetition rate of up to several million per second has led to the use of some form of binary-coded data in the majority of modern high-speed computers. The exact form of coding actually used varies greatly from machine to machine and often more than one form may be used in different parts of one machine: as illustration, two types of widely used coding will be described.

Straight Binary Representation

The straight binary equivalent of a decimal number may be obtained in a variety of ways. One simple method consists in repeatedly dividing the decimal number by 2 and indicating the remainder at each stage by a "0" or a "1" as the case may be. For example the binary equivalent of the decimal number 75 is obtained as follows:—

75	37	18	9	4	2	1	0
	1	1	0	1	0	0	1

The binary number is given by the reverse of the remainder column, i.e. 1001011. The individual digits, read from right to left, indicate whether or not successive powers of 2, i.e. 2^0, 2^1, 2^2, 2^3, etc. are present or absent in the number.

The conversion from binary to decimal notation can be made by adding together the powers of 2 present in the binary number, e.g. in the above example

$$\begin{aligned} 1001011 = \quad & 1\times 2^0 = 1 \\ & +1\times 2^1 = 2 \\ & +1\times 2^3 = 8 \\ & +1\times 2^6 = 64 \\ & \overline{\quad 75} \end{aligned}$$

It will be observed that, in this example, seven binary digits are required for the representation of a two-digit decimal number; in general just over three times as many digits are needed. The straight binary system is used in the majority of computers intended primarily for scientific work.

Binary-coded Decimal Representation

In one form of this coding the individual decimal digits are represented by their equivalent straight binary code, four binary places being used for each digit, i.e.

Decimal digit:									
9	8	7	6	5	4	3	2	1	0
Binary code:									
1001	1000	0111	0110	0101	0100	0011	0010	0001	0000

so that the decimal number 75 would appear in this form of coding as an 8-digit binary number 01110101. Here it will be observed that four times as many digits are required for the binary-coded form as compared with the equivalent decimal number, hence more storage space is needed inside the computer than is the case with the straight binary coding. Moreover, more time or equipment is required for the internal transfer of information with the machine and for the carrying-out of the various arithmetical processes in the arithmetical unit.

The advantage of binary-coded decimal working is the greater ease with which conversions may be made with the decimal form. Such conversions are, of course, carried out automatically by the machines themselves during the input or output processes but, in general, the simplicity or complication of the conversion process will be reflected in the speed of the operation or the complexity of the equipment. The majority of American and some British computers designed for business use employ binary-coded decimal notation.

Serial and Parallel Machines

A further difference is between serial and parallel machines. In serial machines, numbers and orders are represented as a time-series of pulses on one wire, the position in time of each pulse determining its numerical significance. The least significant digit always occurs first in time to make possible a "carry" to the next more significant digit during arithmetical operations.

A parallel machine has all the digits of a number represented at the same instant of time, each digit on a separate wire. A combination of serial and parallel representation is sometimes used. For example, a binary-coded decimal number could have each digit represented by four binary digits in parallel, while the decimal digits themselves occur in series on the four wires.

In general parallel machines are faster than serial machines but need more equipment, as an adding circuit is necessary for every digit, whereas one is sufficient for all digits in a serial machine.

Fixed Point and Floating Point Machines

Numbers are usually specified by between 6 and 12 decimal digits or by the equivalent (between 20 and 40) of binary digits. In a computer working *fixed point*, the decimal or binary point is always considered to be in the same position in all numbers. In a computer working *floating point* the number is specified as $x.2^y$ where x is a number specified by some of the total number of digits and y an exponent specified by the rest of the digits. Floating point representation complicates the design of a computer, but the greater range of specifiable numbers makes it a much more powerful machine.

Form of Order

The form of order used is another important characteristic. It is usually described by the number of addresses or storage locations specified in the order. A one-address machine has orders of the form: "Add the contents of store address X to the accumulator", the accumulator being a single number store or register attached to the arithmetical unit. A two-address machine has orders of the form: "Add the contents of store address X to that of store address Y and put the result in store address Y." In such a machine the contents of store X would be left intact, whereas the new contents of store Y would destroy the original information. In a three-address machine the result would be put in store address Z. Successive orders are taken from consecutive store addresses.

A special two-address machine, often referred to as a one plus one-address machine, has orders similar to a one-address machine with the addition of the address of the next order to be obeyed. By careful choice of the addresses of both orders and numbers this type of order enables the effective access time of the store to be made much less than would otherwise be the case, giving optimum programming.

An important facility built into most digital computers allows the computer itself to modify orders taken from the store before they are obeyed. This is done by the provision of a group of special registers, the contents of a specified one being added to the order before it is obeyed. By this means a single order may be used to do many different operations by altering the contents of the modifying register between each operation. The group of modifying registers is sometimes called the *B-store*.

The Store

The type of store largely determines the form taken by the rest of the machine and also its overall speed. Some stores are essentially serial in operation while others are parallel, and this will usually determine whether the arithmetical unit also is serial or parallel. One of the most important speed-determining features of a store is its access time; that is, the time it takes to obtain access to a specified store address.

Early Types

Early computers used electro-magnetic relays and thermionic-valve flip-flop circuits as stores but these were slow and expensive. One of the first successful stores was of a delay-line. A tube filled with mercury has a quartz crystal mounted at each end, Fig. 2 (a). Electrical pulses applied to one crystal cause waves to travel down the mercury and when they arrive at the other crystal a small voltage is generated across it. This voltage is amplified and fed back into the input crystal causing the pulse to circulate. A typical delay line has a tube 5 ft. long

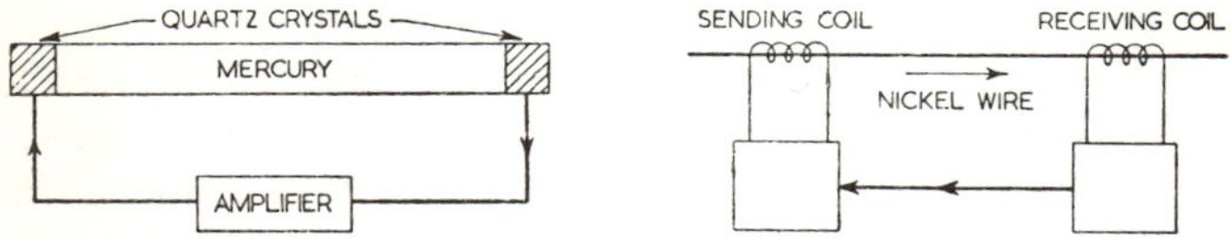

(a) *(b)*

Fig. 2.—*Delay-line stores.*

with a delay of 1 ms, storing 500 pulses at 2 μs intervals. The average access time for such a store would be 500 μs, half the total delay.

Another form of delay-line store uses the magnetostrictive properties of nickel. If a current pulse is passed through a coil wound around a nickel wire, Fig. 2 (b), magnetostrictive effects in the nickel cause a stress wave to travel along the wire which can induce a small current in another coil spaced at a suitable distance, the delay being approximately 2 μs/cm.

A common form of storage is the magnetic drum. A cylinder plated with nickel or sprayed with a layer of magnetic oxide is mounted on bearings so that it can rotate. Recording heads are placed about 0·025 mm from the surface of the drum. By passing pulses of current through a winding on the head, flux is induced in the gap and information is recorded on the drum surface as a series of magnetised spots. As these spots pass under the reading head (which may be the same as was used for writing) they induce small voltages, which are amplified to enable information to be read from the store. The average access time of a magnetic drum store is half the time of rotation, a typical value being 10 ms for a drum rotating at

3,000 rev/min. A medium-size drum stores about 250,000 binary digits on a cylinder of 15 cm diameter and 20 cm long.

The first successful quick-action store used an ordinary cathode-ray tube. Information is stored as areas of charge on the screen of the tube, arranged in the form of a raster. A single 5 in. cathode-ray tube has been used to store 1,280 binary digits.

Magnetic Core Store

The electrostatic store has been superseded in computers by the magnetic core store, using magnetic materials with almost square B/H loops. The binary digits 1 and 0 are represented

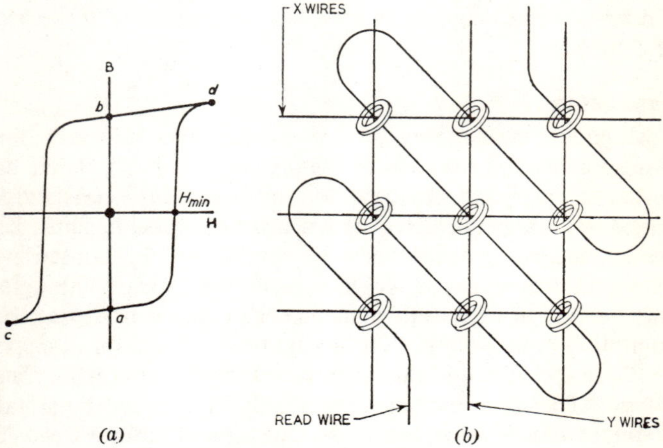

Fig. 3.—*Magnetic core store.*

by the remanent states *a* and *b* respectively, Fig. 3 (a). To read out of the store, a field is applied to take the material to *d*. The presence of a "0" or a "1" is detected by the difference of flux change between the excursion $b \rightarrow d$ and $a \rightarrow d$. The store usually consists of small rings or toroids of ferrite, threaded by several wires. A typical arrangement is shown in Fig. 3 (b). A particular core is selected by passing a current pulse producing a field somewhat less than H_{min} down both the X and the Y wire threaded through that core. The field produced by a single wire is insufficient to change the state of any core, but where the wires intersect twice that field is produced and takes that core to *d*. The original state of the core is sensed by the voltage induced on a "read wire" threading all cores.

Reading from a core destroys the information stored in it and therefore a reading operation is always followed by a re-writing operation if it is desired to retain the information in the core. Writing is similar to reading, but with the direction of current in the X and Y wires reversed. Stores of this type holding around 100,000 binary digits are common, and the trend is towards larger ones. The cost per stored digit is several times that for a magnetic drum. The access time is of the order of 1 to 4 μs.

Tape Stores

Magnetic tape stores are often used when a very large storage capacity is required. The tape usually has a plastic base with a layer of magnetic oxide on one side. Tape consisting of a non-ferrous metal plated with a layer of nickel has also been used. Recording techniques are similar to those used with magnetic drums, and a single reel of ½ in. tape, 2,400 ft. long with eight parallel information channels can store some 20 million binary digits. The access time is long, for at a speed of 100 in./sec it would take just over 4 min to run through such a tape. This disadvantage can usually be reduced by suitable arrangement of information on the tape, organisation of the problem and programming so that information is called up before it is required.

Arithmetical and Control Units

The arithmetical unit performs all the actual arithmetic done in the computer. It is usually designed to add, subtract and multiply two numbers when so ordered. (Some computers have arithmetical units which divide as well.) Usually considered as part of the arithmetical unit are one or more registers each capable of storing one number. The use of the accumulator register in one-address machines has already been described. A multiplicand register is also often provided and the multiply order forms the product of the contents of the multiplicand register and a store address and adds it into the accumulator.

The design of the arithmetical unit is based on adder and/or subtractor circuits. Multiplication and division, if provided, are usually done by repeated additions or subtractions. The whole unit is built up from many interconnected, but in themselves simple, two-state circuits, gates, amplifiers, etc. They are often referred to as logical elements and their interconnection is the logical design of the unit.

Originally logical elements were at first built up from electro-magnetic relays, later from thermionic valves. The transistor and magnetic core now predominate.

The control unit is built up from the same basic logical elements as the arithmetical unit. It contains some form of control register which stores the address of the next order although in the one plus one-address type of computer this is combined with the instruction register which stores the order being obeyed.

Input and Output

Many different methods and media are used to take information into and out of a computer. Apart from a manual input direct from some form of keyboard, one of the earliest forms used punched cards or tape, e.g. 5-hole teleprinter tape. Standard card readers and punches are modified to give electrical outputs and to receive electrical signals from the computer. Output is punched tape from a tape perforator or direct printing on a teleprinter, both equipments receiving electrical signals from the computer.

Paper tape readers are available which operate at 400 up to at least 1,000 characters/sec. Punches will punch up to 250 characters/sec, and, although standard teleprinters are limited to 7 characters/sec, high-speed line printers are available to print 15 lines/sec each of 120 characters.

In almost all cases input numbers are in decimal form, being converted to the binary or other code used by special input circuits or by the arithmetical unit itself. Output is usually decimal, too, conversion back being done in a similar way.

Mechanical input and output equipment may limit the speed of the computer. Magnetic tape can give fast input and output at speeds of around 10,000 characters/sec. Separate machines enable information to be recorded directly on the magnetic tape from a keyboard or from punched tape or cards. Other machines accept magnetic tape from the computer and feed direct to high-speed printers and punches.

Computer Circuits Using Transistors

In the last few years, transistors with very high operating speeds have become available so that the transistor has come to replace the valve in computers. Most of the basic techniques

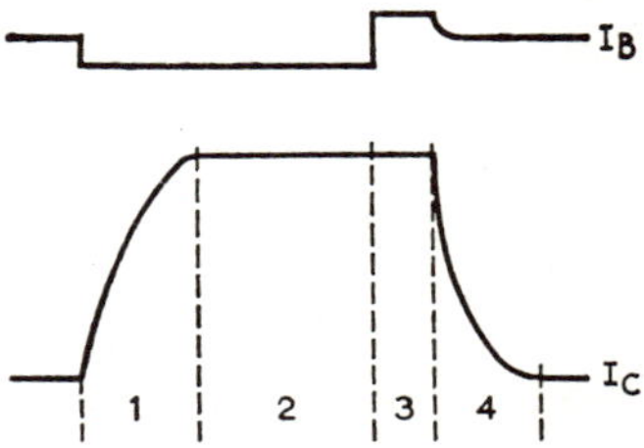

Fig. 4.—*Pulse response of a grounded emitter amplifier.*

used in thermionic valve computing circuits can be adapted to employ transistors; in addition, a number of techniques have been devised which make use of special features of transistor characteristics. Transistors have a number of important advantages over valves in digital computers, including lower power consumption and small size. The ability of the transistor to operate at low signal levels is a particular advantage because the transition time between the levels representing the two digit states can be made much shorter.

When the transistor is used as a computing element it is required to operate with two well-defined states. In contrast to valves, the transistor base has a low input impedance and must be supplied with current. The behaviour of the transistor when switched at high speeds is of special importance in the design of computing circuits. The behaviour of a grounded emitter circuit when switched is shown in Fig. 4. In region 1 the current increases at a rate largely dependent on the ratio of ultimate collector current to input (base) current and the frequency dependence of β (an explanation of β is given on page 68). In region 2 the transistor is "bottomed" (see page 103). When the transistor is turned off, it remains in the low impedance, bottomed state for a short period; this is due to the hole storage (region 3). The active fall (region 4) begins after a certain amount of charge has been supplied by the base circuit to clear the holes from the base region. The hole-storage effect also appears in semiconductor diodes.

The transistor version of the Eccles-Jordan circuit is a basic one in computer technology. Consider the state when the

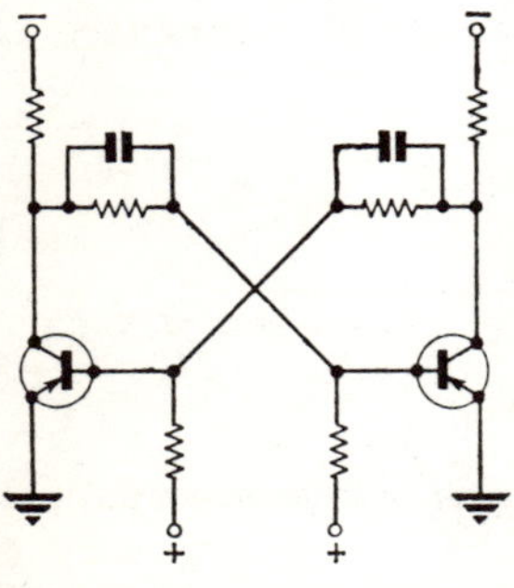

Fig. 5.—*Transistor Eccles-Jordan bi-stable circuit.*

left-hand transistor is conducting (Fig. 5). The collector is bottomed and will be a fraction of a volt (say 0·1) below earth level. The d.c. coupling chain will then maintain the base of the right-hand transistor above earth and this transistor will be cut off. The collector of the right-hand transistor will be negative and current will accordingly flow from the base of the left-hand transistor, maintaining it in the bottomed condition. A pulse applied to the circuit (see pages 133–134) will reverse these conditions.

AND and OR Circuits (see also Section 9)

Transistors are particularly suitable for use as AND and OR switches. Fig. 6 shows an arrangement of three emitter-followers with the emitters connected together. This will

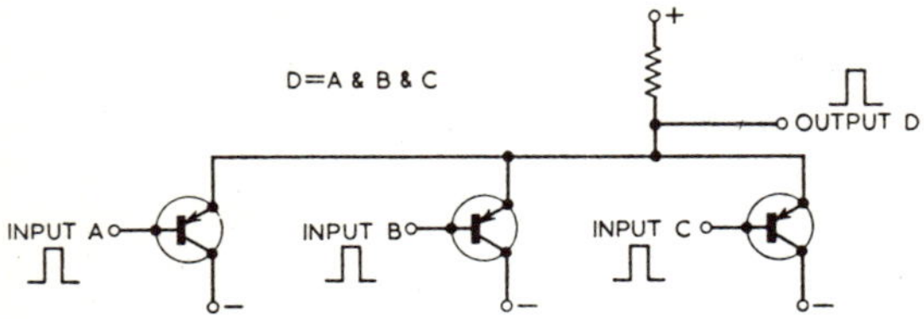

Fig. 6.—*Transistor* AND *circuit* (*for positive pulses*).

operate as an AND gate for positive pulses. The emitter and base may be regarded as the anode and cathode respectively of a diode, with the important distinction that the bulk of the emitter current flows to the collector and not to the base. The

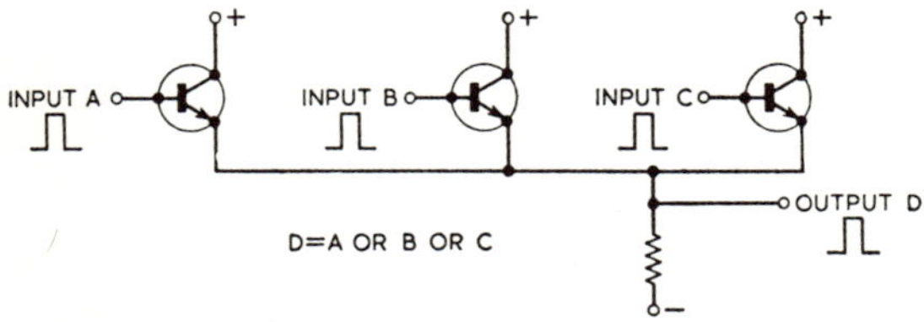

Fig. 7.—*Transistor* OR *circuit* (*for positive pulses*).

transistor circuit thus has a higher input impedance than the diode arrangement. At very high frequencies, however, the advantage is not so great. The corresponding OR circuit is shown in Fig. 7. This makes use of npn transistors. With these the emitter current is cut off by making the base negative with respect to the emitter, and the current is turned on by making the base positive. Thus when any of the bases go positive the common emitters will rise.

The very small potential between emitter and collector when the transistor is bottomed permits the use of direct coupling between stages. The collector of the first transistor of Fig. 8 will be very near earth potential when it is on. The base of the second transistor can be directly coupled to it, the potential being sufficient to keep the current in the second transistor at a low value. The resulting circuits make use of only transistors and resistors. An alternative form of direct coupling is shown

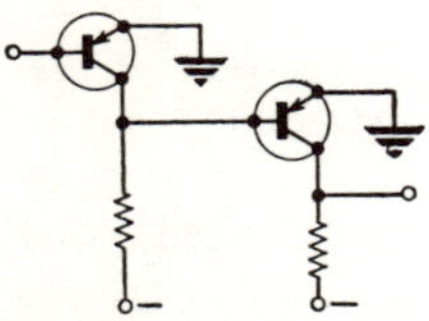

Fig. 8.—*Direct-coupled transistors.*

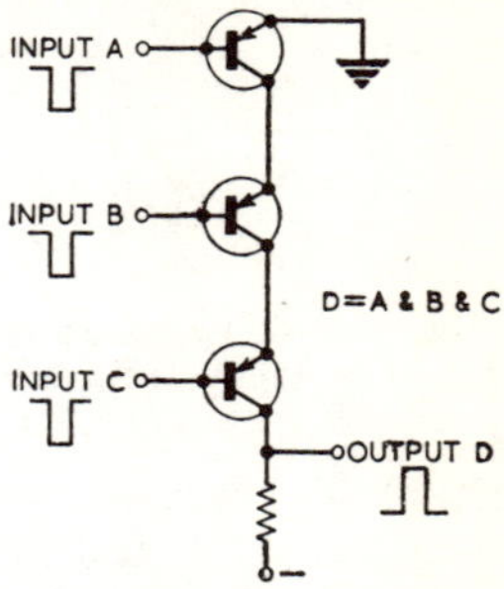

Fig. 9.—*Direct-coupled* AND *gate.*

in the AND gate circuit shown in Fig. 9. Here each transistor operates as a switch, the "contacts" being formed by the emitter and collector. The switches are closed by making the base negative with respect to the emitter. When negative pulses appear on each input lead concurrently, there will be a low impedance path through each transistor and the output potential will rise nearly to earth level.

Magnetic Core-Transistor Circuits

A range of circuits uses a combination of square loop magnetic cores and transistors. A typical arrangement of windings on a core is shown in Fig. 10. The core will change from one remanent state to another when there are sufficient net ampere-turns in the required direction. The windings are arranged so that a standard current of magnitude I in A or B will produce a magnetising force sufficient to change the core from one state to the other. The number of turns and direction of winding C are arranged in opposition to A and B so that with a current I in C the change of state can only occur with a current I in both A and B.

A system of computing circuits can be made up using magnetic cores alone. Amplification is obtained by making use of the flux-resetting or Ramey amplifier principle (see

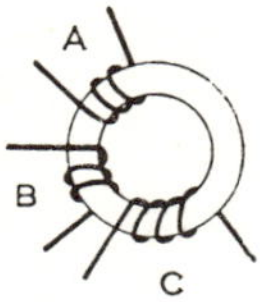

Fig. 10.—*Typical arrangement of windings on a core.*

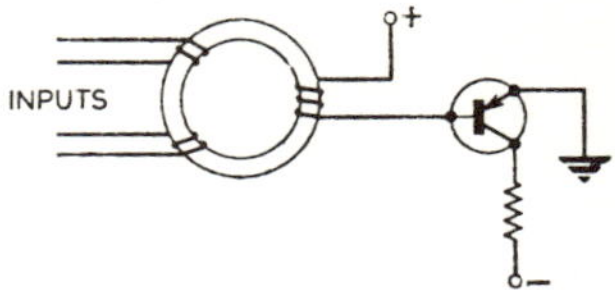

Fig. 11.—*Basic core and transistor circuit.*

Section 9). Alternatively the cores can be used solely as logical elements and amplification obtained with transistors. The basic circuit is shown in Fig. 11. In the quiescent state the transistor is held in the "off" condition by the bias voltage.

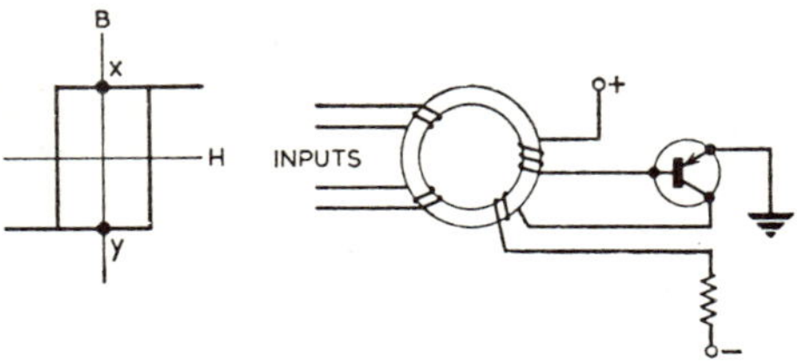

Fig. 12.—*Regenerative core and transistor circuit.*

The magnetic core drives the base potential negative, turning the transistor on, when the state of magnetisation is changed from that representing a one to that representing a zero. The circuit is arranged so that there is sufficient drive to bottom the transistor, and the output current is defined independently of the transistor characteristics. The magnitude of the output current is sufficient to operate further similar circuits. A variant on this is shown in Fig. 12. When the core is in either remanent state (X or Y), the incremental permeability of the core is very low and the core is a poor transformer. When the core is switched from the one to the zero state, the incremental permeability increases considerably and the feedback produces a regenerative action which continues until the switching is complete. The advantage of this arrangement is that a change in the state can be initiated by a very short pulse and the change will be completed without further dependence on the input waveform. A wide range of computing circuits can be produced using this technique, and these have the advantage of being relatively unaffected by temperature. The transistor is only required to supply a large current for a short period, and the small temperature dependent current which flows when the transistor is "off" has no influence on the operation of the circuit. The circuits do not appear to be capable of very high-speed operation, and a pulse repetition frequency of 100 kHz is typical.

ANALOGUE COMPUTERS

In analogue machines, numbers are represented by some physical quantity such as length, angular rotation or electrical voltage, which can be set for the input of data and measured for output. The principle of operation of the analogue

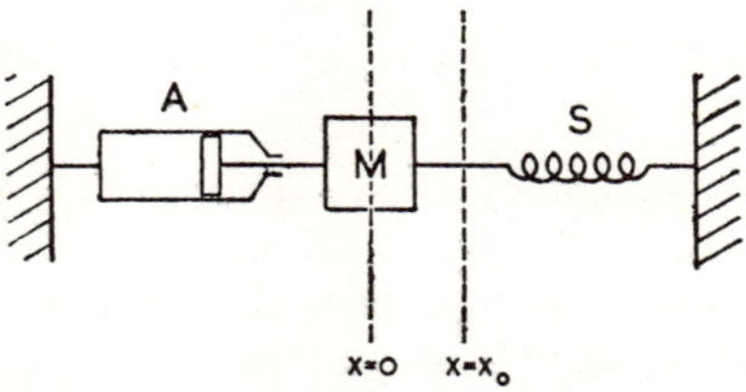

Fig. 13.—*Simple example of the principle of operation of an analogue computer.*

computer may be demonstrated by considering its application to the solution of a simple differential equation arising from the problem shown in Fig. 13. M is a mass fixed on the one hand to spring S and on the other to damper A. If the mass M

is constrained from the neutral position and released at time $t = 0$, the equation describing its motion at any instant $t > 0$ is

$$M\frac{d^2X}{dt^2}+A\frac{dX}{dt}+SX=0 \qquad (1)$$

which gives

$$M\frac{d^2X}{dt^2}=-A\frac{dX}{dt}-SX. \qquad (2)$$

If all these terms are represented by voltages varying with the time, the equation may be solved by the feedback system shown in Fig. 14 using a series of operational amplifiers (1–4).

This diagram is derived by assuming the quantities X and dX/dt already exist. These voltages are scaled according to the

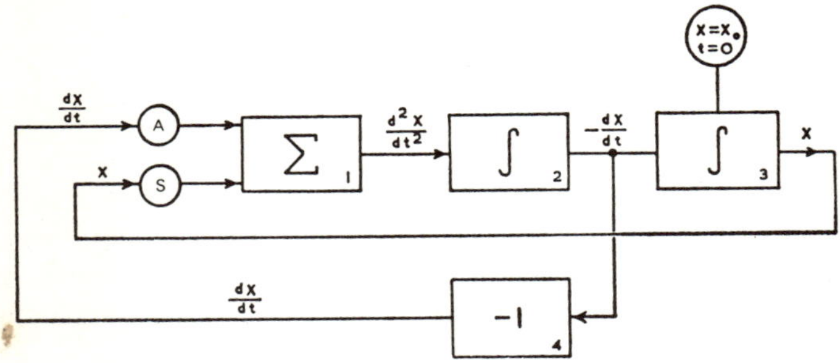

Fig. 14.—*Feedback system used to solve the equation describing the problem shown in Fig.* 13.

appropriate coefficients in the equation by means of the attenuators S and A. These quantities are summed in unit 1 to give d^2X/dt^2, and then integrated with respect to time in unit 2 to give $-dX/dt$. A second integration in unit 3 gives X. Unit 4 is included to preserve the sign relationship of dX/dt given in equation (2). The outputs X and dX/dt are then connected to the attenuators S and A respectively, to close the loop.

If at time $t < 0$ the output voltages of units 1, 2 and 4 are held at zero, whilst the output of 3 is arranged to correspond to the initial displacement X_0, at $t = 0$ all are simultaneously released and at all times $t > 0$ the outputs of units 1, 2 and 3 represent the acceleration, the velocity, and the position respectively of the mass *M*.

This technique may be readily extended to non-linear problems. For instance, if the spring *S* in the problem

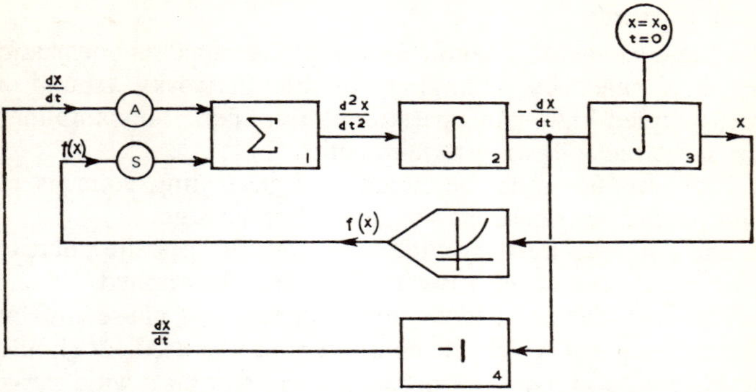

Fig. 15.—*Extension of the system shown in Fig.* 14 *to solve non-linear problems.*

illustrated above has a non-uniform rate displacement characteristic, a function generator representing its variation with displacement may be inserted between units 3 and 1 as shown in Fig. 15.

Computing Elements Required

It is in general desirable for a computer to be suitable for a wide variety of operations. Accordingly, it is customary to perform the more complicated mathematical operations in the computer through combinations of a limited number of simple operations performed by basic computing elements.

A wide range of problems may be solved with the following range of computer elements. (1) Elements that will generate the sum of a number of computer voltages. (2) Elements that will multiply a computer voltage by a fixed co-efficient. (3) Elements that will multiply two or more variables. (4) Elements that will generate functions of machine variables. (5) Elements that will generate the time integral or time derivative of a computer voltage. (6) Delay elements capable of shifting a time variable input along the time axis, without changing its waveform.

Linear Elements

Of these elements (1) and (5) may be described as linear and may be formed by arranging suitable networks around a direct-coupled high gain operational amplifier. This amplifier must meet the following requirements.

(1) The amplifier must be capable of generating voltages of either polarity to represent variables of either sign.

(2) The amplifier must provide an output of opposite polarity to the input, so that negative feedback may be applied.

(3) The amplifier must have high gain and low phase shift at the frequencies at which the computer is to be used. It should also have a good gain and phase margin, so that a wide range of feedback networks may be applied without the risk of instability. The useful values of gain range from 10,000 to about 100,000 at low frequencies. This high gain (increased many-fold by suitable drift correction techniques) increases both the static and dynamic accuracy of the computer and reduces considerably the effect of changing electronic valve characteristics. The phase shift with negative feedback applied depends upon the total internal amplifier gain at computing frequencies; the phase shift at these frequencies, with negative feedback applied so that the overall gain is one, may range from 0·001 to 0·01 degree per cycle per second. The number of cycles of a solution that may be accurately obtained depends directly on this phase shift. Suitable design permits the study of problems in real time, having natural frequencies of up to several hundred cycles per second.

(4) The amplifier must present a high input impedance. The input current must be very small compared with the signal currents in the feedback networks which may be less than 10^{-8} A. Usual input currents range between 10^{-10} and 10^{-12} A on the better operational amplifiers. High values of

input current impose a limitation on the duration of run in computations where integrators are involved.

(5) The spurious signal d.c. at the output must be negligible when no signal is applied to the input. This is known as drift

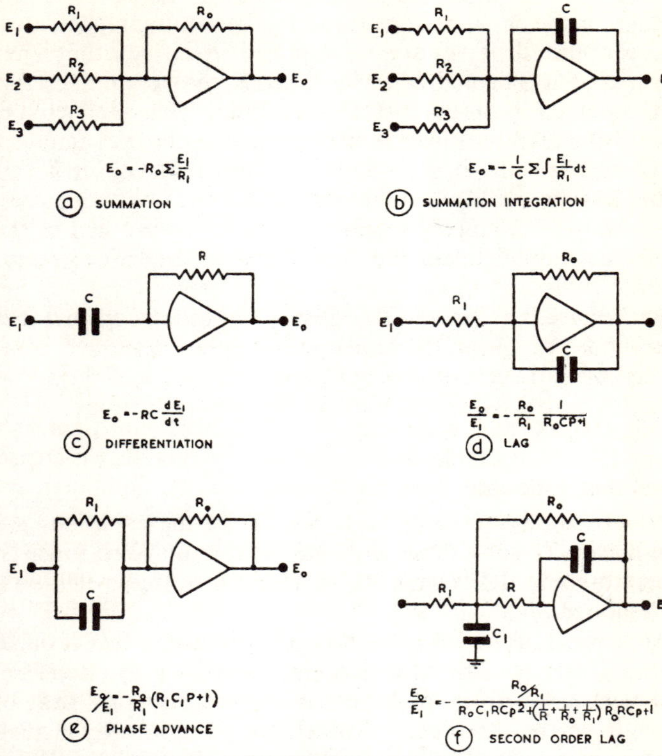

Fig. 16.—*Feedback components connected around a high-gain amplifier to enable various simple linear computing operations to be performed.*

and may be regarded as being due to a small fictitious generator at the input terminal. It arises from a variety of sources, and can be reduced by careful design to less than 1 mV over the period of an hour. By use of an auxiliary drift corrector circuit, this error may be reduced to below 10 μV over a period of days. Such circuits generally consist of a drift-free modulator—a polarised relay working as a chopper at a frequency between 50 and 100 Hz being a popular choice—which samples the voltage at the input terminal due to the fictitious generator. The chopped drift voltage is amplified in an a.c. amplifier, whose output is demodulated by a second contact on the relay, and smoothed by means of a low-pass filter. The amplified d.c. error is then fed back into a second input of the d.c. amplifier to compensate the drift voltage. Besides reducing drift this circuit has the added advantage that, at d.c. and low frequencies, the effective d.c. amplifier gain is equal to the product of the actual d.c. amplifier gain and the d.c. gain of the drift corrector circuit.

By connecting various feedback components around the operational amplifier, a number of simple alternative computing operations may be performed. These are shown in Fig. 16 together with their associated transfer functions.

Non-linear Elements

A vast number of problems exist where the equations are non-linear. To solve these problems fully, multipliers, function generators and delay elements, in addition to linear elements, are required.

Multipliers, devices for obtaining the product of two variable voltages, may be either electromechanical or wholly electronic. The electromechanical multiplier usually consists of a gang of potentiometers, the wipers of which are positioned by a small electric motor proportional to one variable, whilst voltages proportional to the other variables are applied across the winding of each potentiometer. Ganged potentiometers enable several independent variables to be multiplied by this one variable. The accuracy of this type of multiplier ranges from 0·01 to 1·0 per cent depending on the linearity of the potentiometers. The limited speed of response causes decreased accuracy as the frequency of the input rises above a few cycles per second.

Electronic multipliers are more complex devices which generally form their product by operating on a high frequency carrier of suitable waveform. They are capable of operating with equivalent accuracy over a frequency range that extends from 1 to 3 decades higher than the electromechanical multiplier.

The function generator is a device which stores information in the form of curves. These curves may be mathematical functions such as exponentials, sines, cosines, etc., or arbitrary functions derived from empirical data. Function generators give an output $f(x)$ which is the desired function of the input x. Several forms are available; electromechanical units may have tapped, loaded or shaped potentiometers, or a mechanical cam and cam follower drive; electronic units take the form of biased diode networks, giving segments of variable slope and breakpoint, or a masked cathode-ray tube with a photo-electric pick-up and electronic circuitry, which constrains the spot to follow the mask cut to the shape of the desired function. Function generator accuracies may be in the range of 0·1 to 2 per cent, depending on type. Comparator units form another group of elements in this general category. These units may be considered as special-purpose function generators, since they compare two input voltages and produce an output voltage having a discontinuity when the two inputs become equal to a predetermined ratio. Comparators comprise d.c. amplifier-diode circuits, or amplifier-relay circuits when more than one circuit must be switched. Comparators may be made to perform with accuracies of 0·01 to 1 per cent.

A delay element is a device for shifting a time variable input along the time axis without changing its waveform. This unit may be either electromechanical (e.g. a magnetic tape recorder with recording head and read-out head separated by a distance equivalent to a delay time) or entirely electronic. Both methods suffer from some inaccuracy producing distortion of the waveform or spurious outputs.

Problem Connection

Methods of connecting computer elements to form the problem depend considerably on the applications of the machine. In general it is necessary to disconnect the elements from the problem wiring in the "standby" condition, so that each may be checked individually for faults. Problem wiring may be permanent in a special-purpose computer where one particular type of problem is being investigated. In general-purpose computers, however, inputs and outputs are brought to a number of sockets, and problem connections are made with linking pieces. Patch panels which connect up with all the input and output sockets simultaneously, and may be wired for a particular problem remote from the computer, produce a saving in computer time when a number of problems are being investigated.

Input and Output Units

Input units are of two types: reference voltage units having accurately stabilised outputs of 1 machine unit, e.g. 100 V, from which all initial condition voltages may be scaled with suitable potentiometers; and forcing function units from which input voltages such as step and ramp, square wave, noise, sine wave and other waveforms may be derived.

Output units are of two general types: devices for measuring steady-state voltages on the machine, and recorders or visual displays for obtaining the dynamic solutions.

Steady-state measuring devices may consist of very accurate potentiometers, of the switched or sliding variety, connected to the machine reference supplies, which may be used with a null indicating meter to give the value of any computer voltage. Units with automatic balancing circuits and digital read out incorporated (digital voltmeters) make rapid and accurate setting of computer voltages and potentiometers possible.

Recorders

Recorders belong to two general categories: one type records a variable voltage with reference to time, the other one variable voltage as a function of another.

Direct writing recorder units plotting a variable as a function of time are capable of recording up to frequencies of the order of 100 Hz with errors as little as 5 per cent. At lower frequencies, errors may be less than 1 per cent. Mirror galvanometer types using ultra-violet sensitive paper may be used up

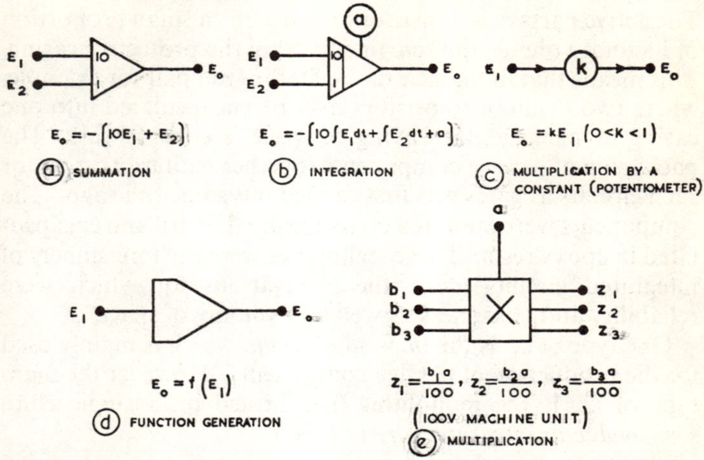

Fig. 17.—*Diagrammatic representation of standard computing elements.*

to 5 kHz. Direct writing X–Y recorders (plotting one variable as a function of another) are considerably slower in response, but more precise. Electronic units using a cathode-ray tube display are also available for both applications. Permanent records may be made photographically. This form of display unit is eminently suitable when a problem is solved repetitively (i.e. problems in which the time scale has been speeded up so that the solution is repeated several times a second) as the effect on the response of gradual changes of problem parameters may be noted.

This completes the list of elements required for the majority of problems. Other, more complex units may be built up from these standard units. Fig. 17 shows in diagrammatic form the standard elements that have been discussed.

11

INTEGRATED CIRCUITS

The active parts of a transistor only occupy a small proportion of its total volume, the remainder being the protective casing. This means that in the case of the Darlington pair for example, where two or more transistors may be encapsulated into one casing or package, the saving of space is considerable. The packaging of passive components together with transistors for such circuits as gates was first carried out some time ago. The components were mounted on to a printed board and encapsulated in epoxy resin. These techniques were the forerunners of integrated circuits (i.c.s), the motivations for which were reliability and cheapness as well as economy of space.

One type of i.c. is the *thick film circuit*, which is mainly used for the production of passive components. However the main type of i.c. is the monolithic (i.e. formed on a single chip) *semiconductor integrated circuit* (s.i.c.).

Manufacture of S.I.C.s

The construction of one npn transistor in an s.i.c. will now be described. However the significant point to remember is that the same process can be used to make many more components simultaneously on the same chip. The transistor is to be formed on to a p-type chip of silicon called the substrate shown in Fig. 1 (a). A layer of n-type silicon is epitaxially grown on top of the substrate (Fig. 1(b)). It is now necessary to isolate the transistor from all other components on the chip. This process should be followed carefully because the formation of the various transistor regions is similar. The top surface of the chip is oxidised to form a layer of silicon dioxide, SiO_2 (Fig. 1 (c)). A photo-etching technique now removes SiO_2 to an accurate pattern so that small islands remain where devices are required (Fig. 1 (d)). P-type impurities are now diffused right down through the n region to the substrate, the function of the SiO_2 being to prevent diffusion in certain parts of the chip. At this stage the chip will be as shown in Fig. 1 (e). The diffusion process which leaves the chip in this state is called isolation diffusion because, with the substrate held at a negative voltage, all pn junctions are reverse-biased and all n regions consequently insulated from each other.

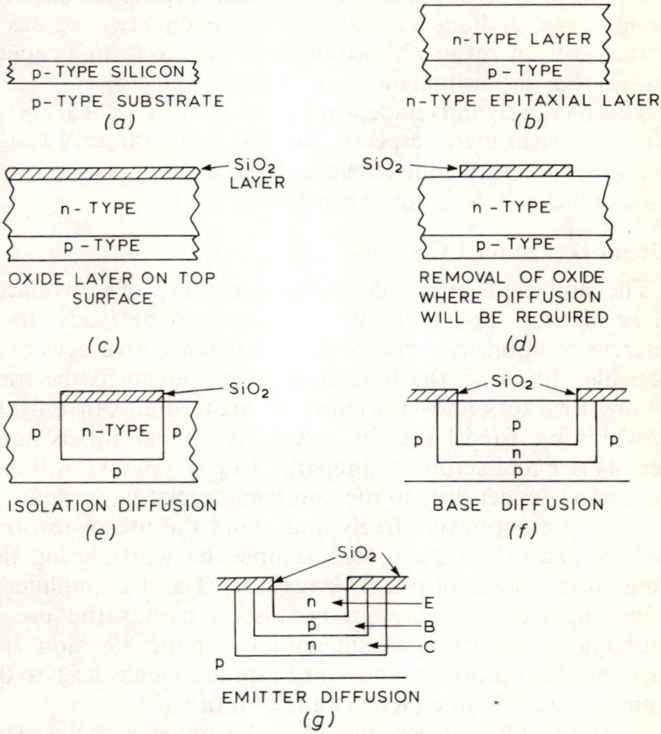

Fig. 1.—*Steps in manufacturing an* npn *transistor in an integrated circuit*.

All that is now necessary to form the transistor is to repeat twice the oxidation, photo-etch, diffusion sequence, first with p-type impurity to form the base (B) as in Fig. 1 (f) and then with n-type impurity to form the emitter (E) as in Fig. 1 (g). The collector (C) is formed by the n-type isolation region.

To complete the i.c. a coating of aluminium is added to a pattern defined by the photo-etching process to form the interconnections between components on the chip. The output leads are then added and the circuit encapsulated.

The same type of process can be used to produce different components. A diode is available by using only two regions of a transistor. A reverse-biased diode is used to form a capacitor. A resistor is simply a piece of semiconductor.

The tolerances and temperature coefficients of s.i.c.s are poor. Much better in these respects, and also in the range of values available, are thin film circuits. Hybrid circuits which use both of these i.c. techniques can also be made.

Circuit Design for I.C.s

The limitations of i.c.s have led to something of a revolution in i.c. design. For economy in circuits to be made up of discrete components, one would use as few active devices as possible. In an i.c. the fundamental component is the transistor. Resistors and capacitors of high value can only be provided by using large slice areas. As a rule, this is not a penalty a manufacturer is prepared to pay since he will lose on packaging density; so the approach to circuit design is to use active components freely and avoid the use of resistors and capacitors. It may, for example, be worth using five transistors instead of one 47 k resistor. The d.c. amplifier is commonplace in i.c. design because it avoids the use of coupling capacitors. The differential amplifier (Section 7) is particularly appropriate since the i.c. process lends itself to the manufacture of well matched pairs of transistors.

Most difficult of all components is the inductor. If the use of inductance is unavoidable, the straightforward solution is to use an external coil. An alternative is to simulate a positive reactance (inductance) by using an active device to invert the negative reactance of a capacitive circuit.

Linear I.C.s

The choice of circuits to be integrated is influenced by their versatility. If a basic circuit has a wide range of uses it can be made in sufficient quantities to be marketed at a low price. Thus a d.c. to 100 MHz amplifier could, by using suitable externally connected components, be employed as an r.f. amplifier, an i.f. amplifier, an audio amplifier, a d.c. amplifier and so on.

Some desirable properties of a typical multi-purpose amplifier would be: (a) high input impedance to prevent overloading the circuit which feeds it; (b) low output impedance to satisfy a variety of load requirements; (c) high gain and wide bandwidth.

One of the simplest general-purpose linear i.c. amplifiers is the Fairchild μA 730 shown in Fig. 2. More complex circuits have been produced for more demanding applications. For example, the gain can be increased by adding a second differential amplifier. The output impedance can be lowered by cascading emitter followers.

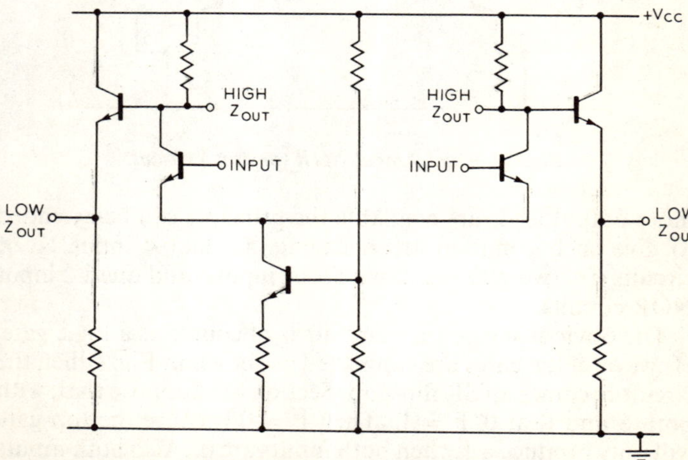

Fig. 2.—*Integrated-circuit amplifier* (*Fairchild μA* 730).

Other types of linear i.c. include the r.f./i.f. amplifier, where high output impedance is necessary to avoid damping the tuned circuits (which have to be connected externally) and the voltage comparator, which gives either a high or low d.c. output level depending on the polarity of the input voltage with respect to some reference voltage. Its function is similar to the Schmitt trigger circuit (Section 8) in that it will convert sine waves to square waves and clean up noisy binary signals.

Digital I.C.s

One of the simplest digital i.c.s is the dual 2 input NOR (positive logic) gate shown in Fig. 3. On a package where

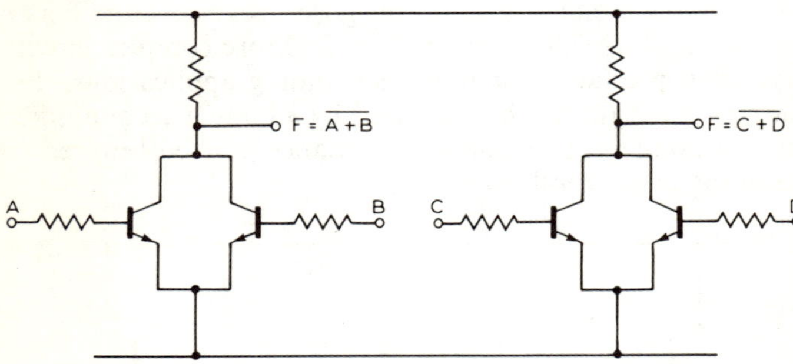

Fig. 3.—*Dual 2 input NOR (positive) circuit.*

more output leads are available the principle can be extended to give such common arrangements as dual 4 input NOR circuits (i.e. two gates each with four inputs) and quad 2 input NOR circuits.

The obvious use of such circuits is of course as a logic gate. However if the gates are connected as shown in Fig. 4 then the circuit becomes an SR flip-flop (Section 8). Suppose that, with both S and R at 0, F = 1. Then $\overline{F}$ = 0 because the top gate will only produce a 1 when both inputs are 0. Also both inputs to the bottom gate are 0, which is correct for F = 1, and the

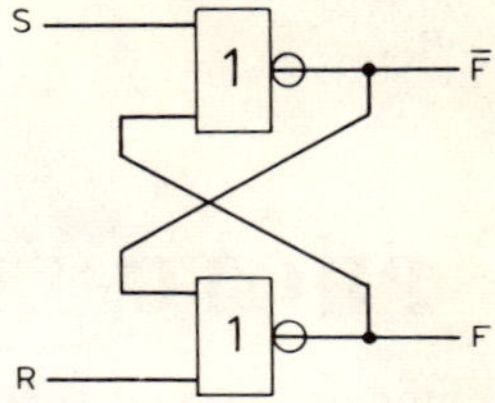

Fig. 4.—*Dual 2 input NOR gates connected as SR flip-flop.*

circuit is stable. A positive pulse at S will have no effect. A positive pulse at R makes F go to 0 which in turn makes $\overline{F} = 1$. This is also a stable state so the circuit is indeed an SR flip-flop. In fact it is quite common to use gates, with external components, for any type of multivibrator since gates of this type are produced in great numbers for computer applications and are therefore cheaper than i.c.s designed specifically as a certain type of multivibrator.

We have already considered (Section 9) the circuits for i.c.s such as D.T.L. and T.T.L. gates and, apart from their basic logic functions, they form the basic elements for such digital i.c.s as flip-flops of all types, memories, shift registers, adders, counters, dividers and so on, all of which are readily available.

Large-scale Integration

Since s.i.c. manufacture involves the production of similar circuits on one large slice of silicon, which is subsequently cut into the much smaller units representing a single i.c., it is possible to retain a large number of circuits in the form of a single multiple circuit system, a technique known as *large-scale integration* (l.s.i.).

12

PHOTO-ELECTRIC DEVICES

Light consists of quanta or small packets of energy. If the light is of short enough wavelength, and so contains energetic quanta, it is possible for the energy in the quanta to be transferred to the electrons in the material and allow them to escape when the material is illuminated with optical radiation. This is called *photo-electric emission.*

ELECTRON MULTIPLIERS

The electron multiplier tube consists of an emissive cathode and a number of secondary emissive surfaces, called dynodes, arrayed as shown in Fig. 1. The primary electron beam from

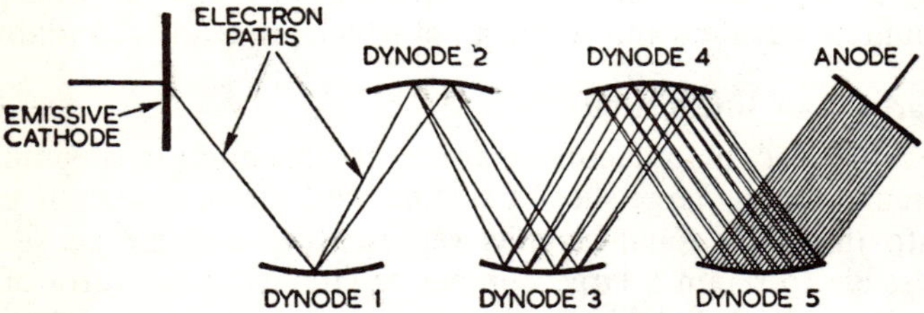

Fig. 1.—*Principle of the electron multiplier. Each successive dynode and the anode are maintained at successively higher positive potentials with respect to the cathode.*

the cathode is focused on to the first dynode. When it strikes this it causes secondary electrons to be given off and directed towards dynode 2, and so on. The cathode emission may be produced in the normal way—by heating—or a photocathode may be used (in which case the tube is known as a photomultiplier). The latter method is most commonly encountered in nucleonics and spectroscopy. The dynodes are made sometimes of silver-magnesium or of an emissive substance similar to the cathode. For every electron striking a dynode, three to five secondary electrons may be emitted. Gains of 80,000 to 250,000 are not unknown. Under the best conditions, using a modern photomultiplier tube, one electron leaving the cathode can produce an output signal that is well above noise level.

In television cameras like the image-orthicon an electron-multiplier stage is an integral part of the camera tube. The photomultiplier tube used as a scintillation counter in nucleonics is a separate tube but must be properly housed because of its high sensitivity. It must be sited in a light-tight box, exposed only to the light of the signal and mounted in such a way that it is not itself a generator of light. For pulse working the photomultiplier is usually contained in an earthed box: the cathode is earthed and the signal fed from the anode through an isolating capacitor. For spectrometry, where a direct measure of the anode current is required, the anode of the tube may be earthed but care must be taken that there are no electrical discharges in the vicinity of the cathode. E.H.T. and signal leads into the photomultiplier housing must have opaque insulating material.

In the photomultiplier, the cathode and its shield form an electrostatic lens focusing the photo-electrons on to the first dynode. For the secondary emission to be at least twice the incident current the voltage applied to the first dynode must exceed 50 V, and for each succeeding stage of multiplication a similar increase in voltage must be applied.

Interstage voltages are usually obtained from a potential divider formed by a series of resistors wired to the tube socket. The dynode chain current should be ten times the mean anode current. Fig. 2 shows a commonly used circuit for scintillation counting. A capacitor between D11 and e.h.t. will improve the characteristics of this circuit for large pulses.

The photomultiplier has a high impedance output (10 MΩ) and is ideally matched into a field-effect transistor with a low input capacity. Under these conditions pulses will be produced, due to single electrons leaving the cathode, which are well above device noise. This is, however, a condition of extreme sensitivity and photo-multipliers are commonly used in probes at the end of several feet of cable.

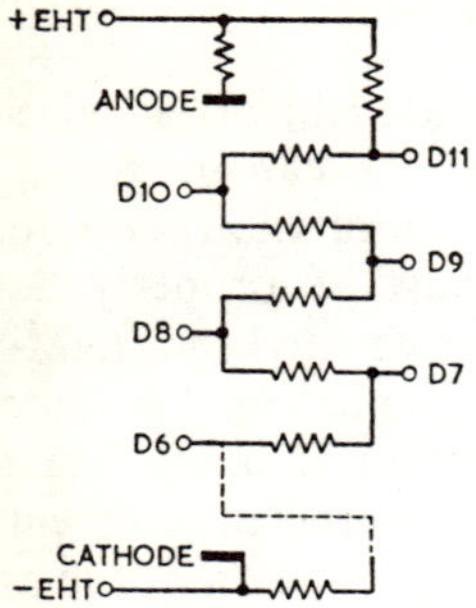

Fig. 2.—*Dynode voltage resistor chain for an electron multiplier tube.*

A photomultiplier in complete darkness has a residual anode current, due to thermal emission and other unwanted processes, which can be expressed as a measurement current or as a count rate. Exposure to light increases the dark current, which will subsequently decay again to its previous value in the dark, provided the maximum safe currents have not been exceeded. For the average photomultiplier a very stable source of e.h.t. is required. Any non-uniform distribution of the dynode voltages will give a lower gain for a given voltage.

Different types of dynode structure can be used in practice. The venetian-blind photomultiplier tube has been developed for scintillation counting applications and ranges in size from 5 cm to 18 cm in diameter. The windows of these tubes are polished flat to provide good contact with the scintillator crystal.

One series of small photomultipliers employs the box-type structure and is used with quartz windows for ultraviolet work. A miniature focused photomultiplier has been developed for use in computer applications.

PHOTO-ELECTRIC CELLS

Certain substances—called photo-emissive—emit electrons when exposed to light. The photo-electric cell consists of a cathode, the surface of which is coated with a photo-emissive substance, and an anode in a tube which may be evacuated or gas-filled. If the anode is made positive with respect to the cathode, the electrons given off by the photocathode when it is exposed to light will flow to the anode. The strength of the current flowing through the cell is dependent, within certain limits, upon the intensity of the light falling on the photo-cathode and the voltage applied to the anode. The cathode may take the form of a semi-cylinder, coated on the inside with photo-emissive material, the anode being a metal rod situated at the axis of the cathode.

The two most commonly used photocathodes are (a) a monatomic layer of cæsium on silver oxide, or (b) a thin layer of cæsium on antimony. The former has maximum

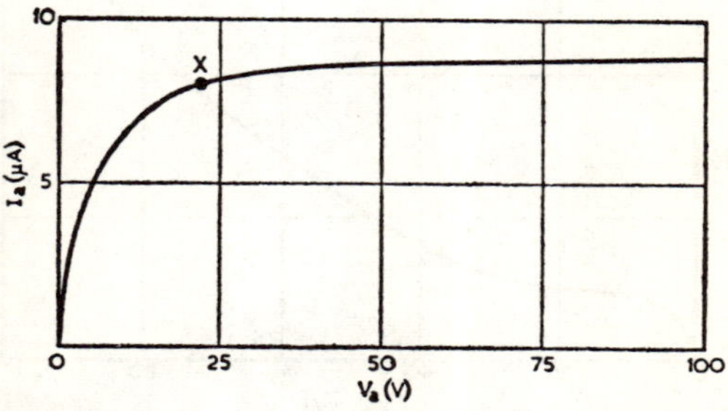

Fig. 3.—*Anode voltage/current characteristic of a high-vacuum photo-cell, for a given anode load and constant illumination.* X *is the saturation point.*

sensitivity to incandescent light and near infra-red radiation, the latter being most sensitive to daylight and to light with a blue predominance.

The I_a/V_a characteristic of a photocell depends upon whether it is a high-vacuum or a gas-filled type. The characteristic of a typical high-vacuum type, for a given light intensity, is shown in Fig. 3. It can be seen that above a certain anode voltage there is little change in anode current, so that at anode voltages above this "saturation level" the anode current is determined entirely by the incident illumination. For a gas-filled photocell, the I_a/V_a characteristic takes the form shown in Fig. 4. In this type of cell the anode current is greater than the current derived from the actual photo-emission because of the gas amplification that takes place, that is the electrons released from the photocathode release further electrons from the gas as they collide with the gas atoms. The degree of gas amplification is largely governed by the anode voltage, the anode current increasing progressively with increasing anode voltage. Because of the gas amplification in a gas-filled photocell, this type of cell is more sensitive than vacuum cells of exactly the same design.

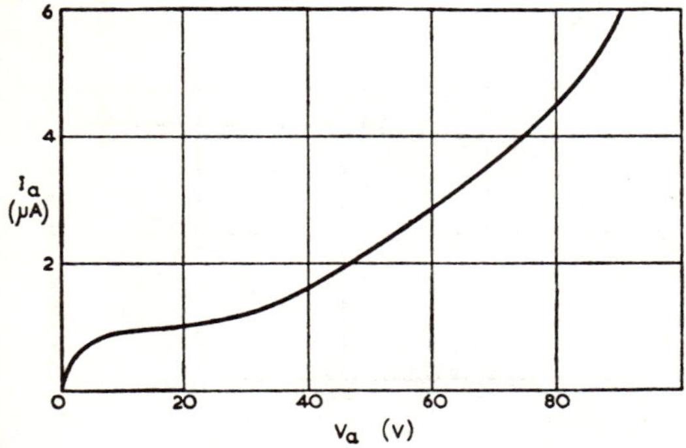

Fig. 4.—*Anode voltage/current characteristic of a gas-filled photocell under constant illumination.*

Provided that the anode voltage of a vacuum cell is kept above the saturation level, the anode current is proportional to the light falling on the photocathode, so that this type of cell is particularly suitable for applications where accurate measurements or proportional control are required. As the anode current of a gas-filled cell varies with anode voltage over the whole of its working range, this type of cell is not suitable for quantitative work, but because of its high sensitivity it has wide application in "on–off" or "stop–go" devices operated by small but sudden changes in illumination.

A typical circuit in which the output of a photocell is amplified by a triode valve and used to operate a relay is shown in Fig. 5. The anode voltage for the photocell is derived from a potentiometer acting as a potential divider. The current flowing in the photocell when it is illuminated develops a potential across resistor R which is applied to the valve for amplification.

A further common type of photocell is based on the photo-conduction or photo-voltaic effect. This makes use of the fact that in certain materials, e.g. selenium, cadmium phosphide, silicon and semiconductor materials generally, light energy increases the number of free electrons present and hence the conductivity of the material, its resistance falling in direct proportion to the intensity of the illumination to which it is exposed.

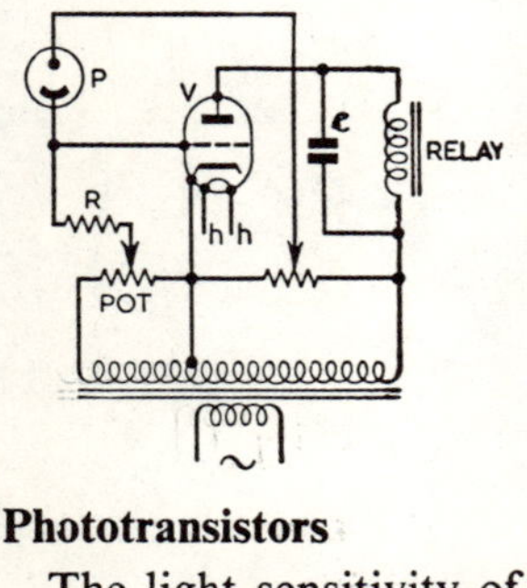

Fig. 5.—*Simple circuit in which the output of a photocell is amplified and used to operate a relay.*

Phototransistors

The light sensitivity of the pn junction in a semiconductor crystal has been exploited to produce a current-generating device with several advantageous properties. The small size and low operating voltages are characteristic of all semi-conductor devices but also these phototransistors have high sensitivity and can be made with very rapid response. Due to the spread of characteristics inherent with transistors this device cannot be used for direct measurement. It does, however, prove most useful for reading punched tapes and cards and for counting the moiré fringes on the diffraction grating measuring scales on automatic machine tools and, of course, for any application where size is important.

Briefly the light falling on a pn junction generates electron-hole pairs and, when the junction is reverse biased, the light-generated current is considerably larger than the minute current passing in a non-illuminated transistor. When the illuminated pn junction is the base/emitter junction, the collector current, amplified by transistor action, is sufficient to work a thyratron. Phototransistors can be made which are sensitive to X-rays, ultra violet and infra red.

13

ELECTRONIC MEASUREMENTS

The instruments described in this section may be classified under three main headings: (*a*) Non-electronic instruments used for fault location and testing of electronic equipment, e.g. multi-range meters. (*b*) Electronic instruments capable of measuring or testing electrical quantities, e.g. cathode-ray oscilloscopes and valve voltmeters. (*c*) Instruments used for measuring non-electrical quantities by electrical methods, e.g. pH meters. It will be more convenient, however, to deal with the various instruments in a somewhat different order depending on the principles involved.

THE MOVING-COIL METER

The moving-coil meter is an essential component of nearly all electrical deflection instruments. In Fig. 1 the light-weight

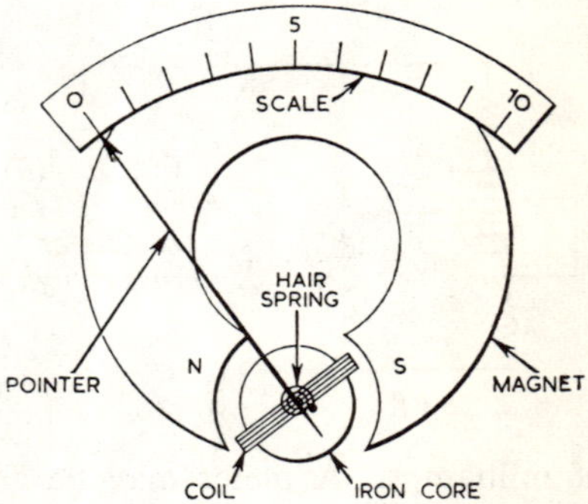

Fig. 1.—*The moving-coil meter.*

coil is pivoted between the poles of the permanent magnet. Two hairsprings (only one shown) retain the pointer, carried by the coil, in line with zero on the scale when no current is flowing. When a direct current is passed through the coil, a magnetic field is produced which acts on the field of the magnet to provide a torque, causing the pointer to move over the scale, under control of the hair springs. An iron core is provided to distribute the magnetic flux radially. This ensures that any increase in current results in a corresponding increase in deflection so that the scale is linear, i.e. evenly divided.

If the current is reversed, the coil will try to deflect the pointer in the opposite direction so that, unless the instrument is of the centre-zero type, the pointer will be driven against the stop. The instrument will give no deflection on alternating current because its inertia will be too great to allow it to follow the current changes.

Extension of Range

Moving-coil meters are normally made to have a basic full-scale deflection ranging from a few microamps to a few tens

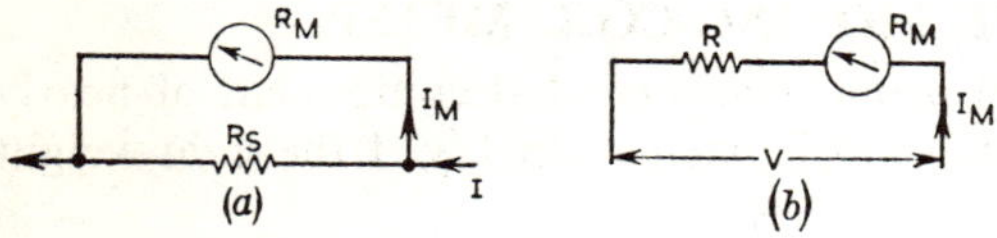

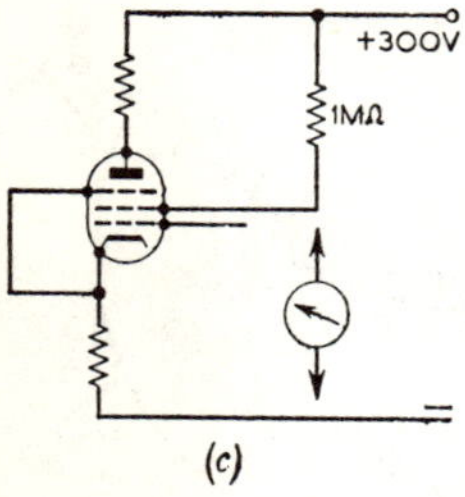

Fig. 2.—(*a*) *Extension of milliammeter range.* (*b*) *Conversion of milliammeter into a voltmeter.* (*c*) *Measurement of voltage across a high resistance circuit.*

of milliamps. A meter may be adapted to measure greater currents by the use of a resistance, known as a shunt, to by-pass the excess current past the coil.

Fig. 2 (a) shows an instrument having a resistance of R_M ohms and a full-scale deflection of I_M milliamps being shunted by a resistor R_S to give an extended range of I milliamps.

$$R_S = \frac{R_M}{(I/I_M)-1} \text{ ohms.}$$

EXAMPLE: *What value shunt resistance is required to convert a* 1 *mA meter to read* 10 *mA if the meter resistance is* 90 *ohms?*

$$R_S = \frac{90}{(10/1)-1} = 10 \text{ ohms.}$$

Ammeter sensitivity.—Ammeters should have low resistance so that their introduction into a circuit in series with existing components has negligible effect on the voltage distribution in the circuit. The resistance of a moving-coil milliammeter or a shunted moving-coil ammeter is usually negligible compared with that of other components in a valve circuit but, in transistor circuits, the introduction of an ammeter may seriously affect circuit operation.

Voltmeters

A moving-coil instrument may be used as a voltmeter by connecting a resistor, known as a multiplier, in series (as shown in Fig. 2 (b)). According to Ohm's Law the current through a resistance is proportional to the voltage across it, hence the scale may be directly calibrated in volts.

$$R = (V/I) - R_M$$

where R is the resistance of the multiplier, R_M is the resistance of the meter, and I is the current at full-scale deflection (f.s.d.).

EXAMPLE: *What value resistor is required to convert a* 50 *microamp meter, having a resistance of* 1,000 *ohms, into a voltmeter with a full-scale deflection of* 100 *volts?*

$$R = 100(10^6/50) - 1{,}000 = 1{,}999{,}000 \text{ ohms.}$$

If the meter resistance had been neglected the error would have been less than 0·1 per cent, i.e. much less than the accuracy with which the scale could be read.

In multi-range instruments different shunts and multipliers are switched into circuit on each range.

Voltmeter sensitivity.—In moving-coil and most other types of voltmeter, the deflection is not necessarily dependent on the voltage across the circuit tested but on the current driven through the resistance of the voltmeter by the voltage at its terminals. When testing high resistance circuits, the current drawn by the voltmeter may seriously affect the distribution of voltage in the circuit. This is shown in Fig. 2 (c). If the screen grid current of the pentode is 0·2 mA the screen voltage will be 100 V. If an attempt is made to measure this by means of a voltmeter having a resistance of 100,000 ohms, an indication of only 30 V will result. This does not mean that the voltmeter is inaccurate, for it may be indicating very precisely the voltage present across its terminals. The voltmeter is just not sensitive enough to provide a full deflection with the small amount of power available from the circuit under test.

The sensitivity of a voltmeter that requires a current of 1 mA for full-scale deflection is expressed as 1,000 ohms/volt. The Avometer Model 8 having a 50 microamp movement has a sensitivity of 20,000 ohms/volt on d.c. ranges.

Rectifier Moving-coil Instruments

A moving-coil milliammeter can be adapted to measure alternating current by first rectifying the current with a bridge-connected rectifier. The deflection is now proportional to the average value of the current but such instruments are normally calibrated to read the r.m.s. value of a sine wave. The resistance of the rectifier varies with the current, particularly at low currents. For this reason, the range of a rectifier moving-coil milliammeter is normally increased by the use of a current transformer—a resistance shunt would produce a non-linear scale. For similar reasons it is not usual to use a rectifier with an instrument having a full-scale deflection less than 1 mA. A rectifier moving-coil milliammeter may be converted into a voltmeter by connecting a resistor in series with the combination. Low voltage ranges require the use of a step-up transformer unless a specially calibrated scale can be tolerated.

VALVE VOLTMETERS

The sensitivity of a voltmeter may be increased by the use of a valve or transistor amplifier between the source of the unknown voltage and the deflection instrument. The amplifier need not increase the unknown voltage but should provide current amplification so that only a much smaller current has to be supplied by the source of the unknown voltage. This means that the resistance across the input terminals of the valve voltmeter is much greater than that of a corresponding non-electronic instrument.

Two types of valve voltmeter are commonly used: (1) D.C. amplifying valve voltmeters. These are normally used to measure d.c. voltages but may be preceded by a rectifier for alternating voltage measurement. (2) Amplifier detectors.

D.C. Amplifying Valve Voltmeters

The basic form of valve voltmeter is shown in Fig. 3 (a). The valve normally passes a steady anode current, which gives a constant voltage drop in the cathode resistor R_k so that the grid/cathode voltage is at point A on the graph shown in Fig. 4. When the input terminals are connected to a source of unknown voltage, the anode current will increase or decrease depending on the polarity of the applied voltage.

The graph shows that no appreciable grid current will flow unless the grid is driven more positive than the cross-over point or contact potential (point C). The input impedance is thus approximately equal to the value of the grid leak resistance R_g, which may be as high as several megohms with some types of valve. It will be seen that below the contact potential (point C) a small negative grid current will flow. This is due to attraction of positively charged gas ions to the negatively charged grid. If measurements have to be made on very high resistance circuits, even this small negative grid current may cause appreciable error. A valve in which the grid current is particularly low (known as an electrometer valve) is useful in these circumstances.

The simple valve voltmeter shown in Fig. 3 (a) provides a deflection dependent on the value of the unknown voltage but,

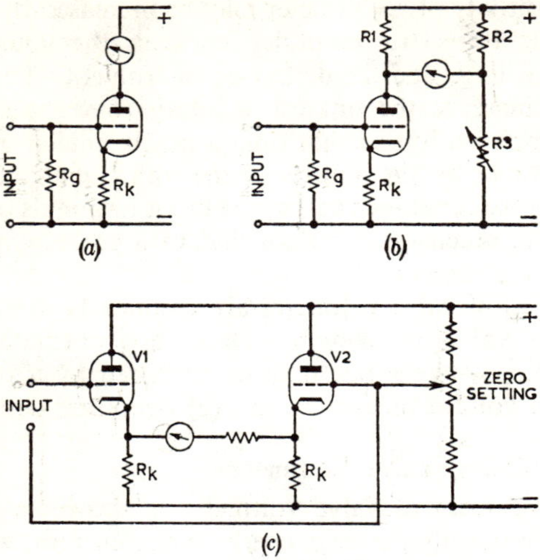

Fig. 3.—*Basic valve voltmeter circuits.*

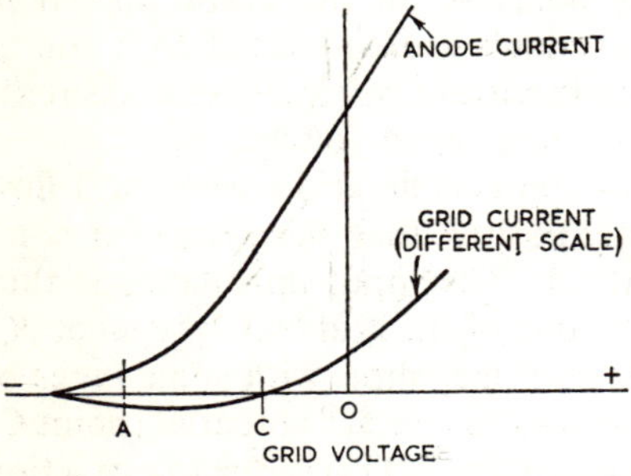

Fig. 4.—*Triode characteristics.*

since the grid current can be very low, it is much more sensitive than a voltmeter made by connecting a resistor in series with the deflecting instrument. Such a simple circuit, however, has the following serious limitations:

(*a*) Because of the standing current in the anode circuit, a deflection will be obtained when no input voltage is applied.

(*b*) A comparatively insensitive meter movement has to be used to carry the standing anode current.

(*c*) The scale is dependent on the valve characteristics and must be specially calibrated.

(*d*) The accuracy is very dependent on changes in valve characteristics and power supply voltages.

Some of these limitations are minimised by the circuit shown in Fig. 3 (b). The three resistors R1, R2, R3 and the effective resistance of the valve may be considered to be a Wheatstone bridge. R3 is first adjusted until the pointer indicates zero with no voltage applied to the input terminals. The bridge is now balanced, hence equal voltages are developed across the valve and R3. When the input terminals are connected to a source of unknown voltage, the voltage dropped across the valve changes, and thus a deflection which is dependent on the value of the unknown voltage is produced across the meter. This circuit overcomes the objections outlined at (*a*) and (*b*) but is still not wholly satisfactory.

If a second valve of the same type is used instead of the resistor R3, the zero setting is much less dependent on power supply voltages.

The bridge circuits so far described have the disadvantage that the "law" of the instrument, i.e. the relationship between deflection and input voltage, is effectively that of the valve. The calibration is thus likely to become inaccurate as the valve characteristics change with age or because of variation in power supply voltages.

In the valve voltmeters so far described, the purpose of the cathode resistor R_k is the same as that of a normal cathode bias resistor in an amplifier, namely to provide a few volts of bias, sufficient to prevent appreciable grid current. If, however, the value of the cathode resistor R_k is increased from a few hundred ohms to say one hundred kilohms, i.e. several times the valve impedance, the performance of the circuit is considerably changed.* When the grid of the V1, Fig. 3 (c) is made positive with respect to that of V2, V1 anode current will increase, but the voltage developed across the cathode resistor will also increase, hence the effective input, i.e. the

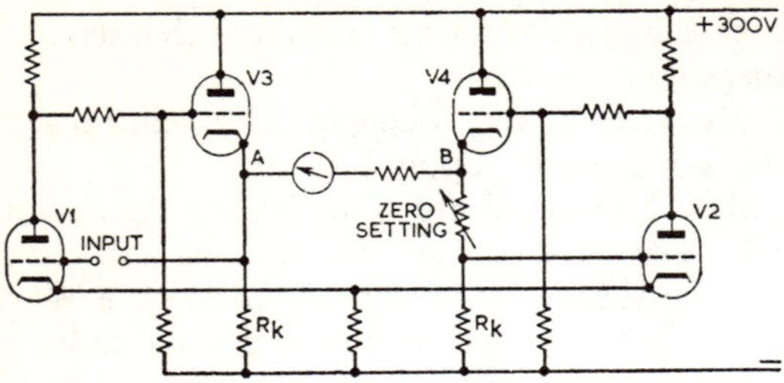

Fig. 5.—*Two-stage d.c. amplifying valve voltmeter.*

voltage between grid and cathode will be much less than the voltage applied to the input terminals. This is a form of negative feedback.

If the meter is now transferred to the cathode circuits, the anode loads can be short circuited. The two valves now become cathode followers, hence the whole of the output is fed back in series with the input terminals.

The voltage gain of each stage is reduced to about 0·9, but the 100 per cent negative feedback results in the effective characteristics being very nearly linear. The input impedance is also increased since the actual voltage between grid and cathode is now less than that applied to the input terminals. It is necessary to transfer the lower input terminal to the grid of V2 to prevent V1 being biased back when a d.c. path is connected across the input terminals.

Fig. 5 shows a simplified circuit of a valve voltmeter developed by Scroggie.† Each half of the circuit comprises a voltage

* *Journal of the British Institution of Radio Engineers*, Vol. VIII, March–April 1948, pp. 86–95.

† *Wireless World*, January 1952, pp. 14–18, and August 1954, pp. 371–376.

amplifying stage V1 or V2, coupled to a cathode follower V3 or V4. Feedback of the whole of the output of V3 via the input terminals to the grid of V1 results in a gain almost exactly equal to 1. This circuit exhibits all the advantages of the double cathode follower circuit shown in Fig. 3 (c), and has the additional advantage of very low output impedance. This means that any voltmeter drawing a current less than a few milliamps and having a full-scale deflection less than about 50 V will, if connected across points A and B, indicate the voltage applied to the input terminals. The current drawn from the circuit connected to the input terminals may, however, be as low as a fraction of a microamp.

Transistor Voltmeters

Transistors are essentially current-controlled devices and hence are not as suitable as valves for use in high impedance valve voltmeters. The characteristics of transistors are also

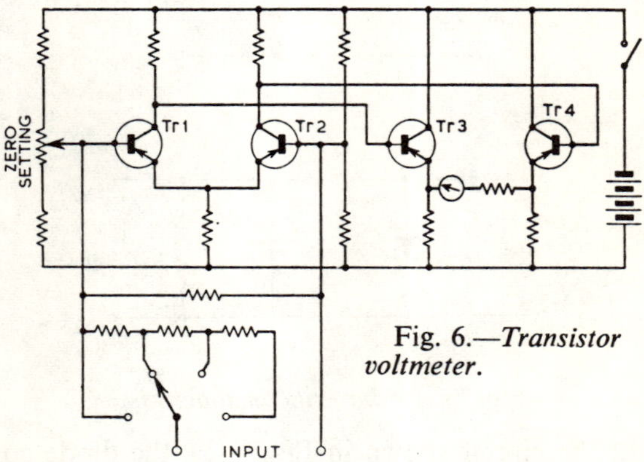

Fig. 6.—*Transistor voltmeter.*

very dependent on temperature. Transistor voltmeters normally have a basic full-scale deflection of only a few millivolts but, by connecting a resistor across the basic input terminals, a number of high value resistors may be switched in series to provide high voltage ranges as in a non-amplifying multi-range voltmeter.

Fig. 6 shows a transistor voltmeter in which transistors Tr1 and Tr2 form a balanced voltage amplifier feeding a pair of emitter follower transistors Tr3, Tr4 which provide additional current amplification. Temperature effects are minimised by the use of balanced pairs of transistors, by the use of silicon transistors and by mounting all the transistors close together in a copper block. Such an instrument may have a sensitivity of 1 megohm/volt.

Diode Voltmeters

The valve voltmeters so far described are only suitable for direct voltage measurement. Alternating voltages must first be rectified before they will provide a deflection on a moving-coil

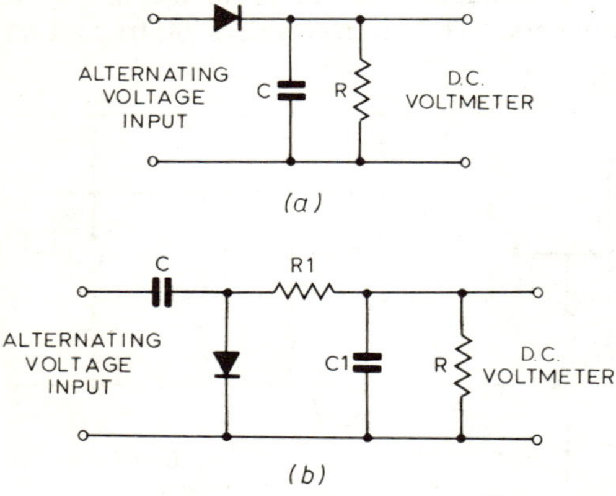

Fig. 7.—*Peak rectifier voltmeters.*

meter. In the circuit shown in Fig. 7 (a), the diode conducts during the first few positive half-cycles of input voltage, charging the capacitor almost to the peak value of the voltage. If the time constant (C × R) is long enough, negligible discharge occurs during the negative half-cycles, hence only a small pulse of current flows during the subsequent positive half-cycles.

Such a circuit is often called a peak rectifier voltmeter, but is usually calibrated to indicate the r.m.s. value of a sine wave. The instrument shown in Fig. 7 (a) is unsuitable for use in circuits not providing a d.c. path between the input terminals or when the alternating voltage being measured is superimposed on a direct voltage. Fig. 7 (b) shows a circuit which does not have these limitations. Further components—R1 and C1—are, however, necessary to isolate the alternating voltage from the output terminals. The d.c. output of a peak rectifier circuit is normally measured on a d.c. amplifying transistor or valve voltmeter.

The Amplifier Detector

The rectification efficiency of a diode becomes greatly reduced at low voltages, hence diode voltmeters cannot normally be used to measure voltages below 0·1 V. In the

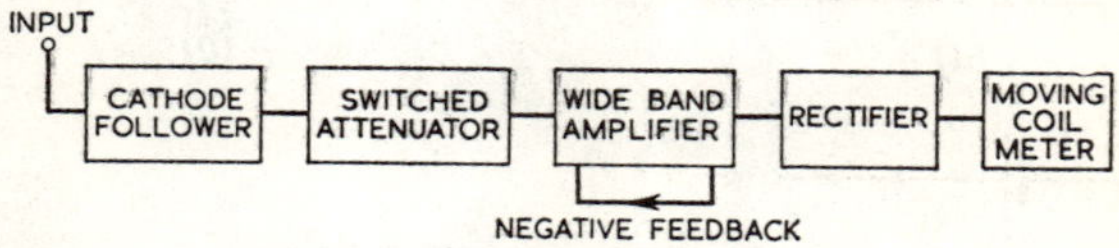

Fig. 8.—*Amplifier detector.*

amplifier detector shown in Fig. 8, a wideband amplifier is used to increase the sensitivity of a rectifier voltmeter. Long-term accuracy is ensured by use of heavy negative feedback. The amount of feedback may be adjustable to provide different voltage ranges or, as shown, a switched attenuator may be used. A cathode follower (often constructed as a probe) is included to increase the input impedance. The amplifier detector is often used as the basis of long-term accurate measurements of audio levels in sound broadcasting.

RESISTANCE AND IMPEDANCE MEASUREMENT

Ohmmeters

The simplest form of resistance-measuring device is the ohmmeter, consisting essentially of a moving-coil meter having a built-in battery in series as shown in Fig. 9 (a). The terminals are first short circuited and variable resistor P adjusted until the instrument deflects to full scale. This point corresponds to zero on the ohms scale. When the short circuit is removed the instrument will show no deflection hence this point becomes infinity or maximum resistance on the ohms scale. Intermediate points are calibrated by connecting known resistors to the terminals. The upper end of the scale is very cramped hence high resistances can only be measured very approximately.

Two adjustments P and R are often provided to correct for variations in voltage and internal resistance of the battery.

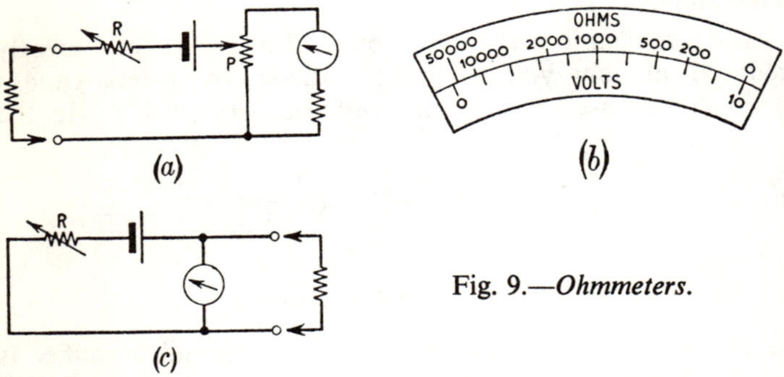

Fig. 9.—*Ohmmeters.*

The range of an ohmmeter can be extended to measure higher resistances by the use of a higher voltage battery and corresponding higher series resistance: lower resistance ranges can be provided by switching in lower resistance shunts across the instrument.

In multi-range instruments the negative side of the battery is usually connected to the positive terminal of the instrument. It is important to remember this when testing polarised components such as rectifiers or electrolytic capacitors. Some multi-range instruments, however, provide a reversing switch to avoid this anomaly.

In another type of ohmmeter, for measuring very low resistance, the unknown resistance is connected across the movement as shown in Fig. 9 (c). The zero point on this resistance scale corresponds to the zero on the current scale. The variable resistor R is adjusted so that the pointer indicates maximum resistance when the terminals are disconnected.

The Wheatstone Bridge

One form of the well-known Wheatstone bridge often used for testing electronic components is shown in Fig. 10. The unknown resistor R is compared with a known "standard"

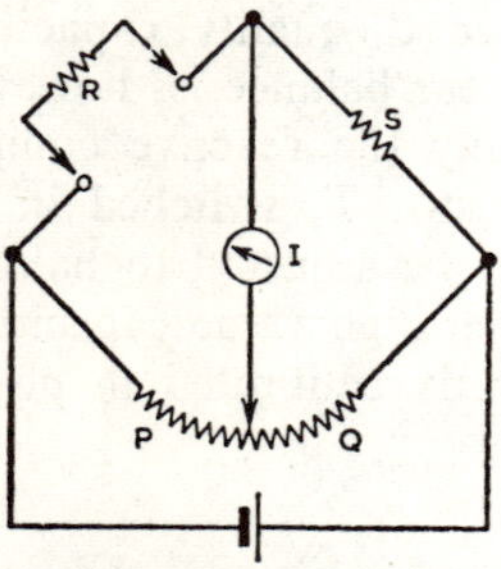

Fig. 10.—*The Wheatstone bridge.*

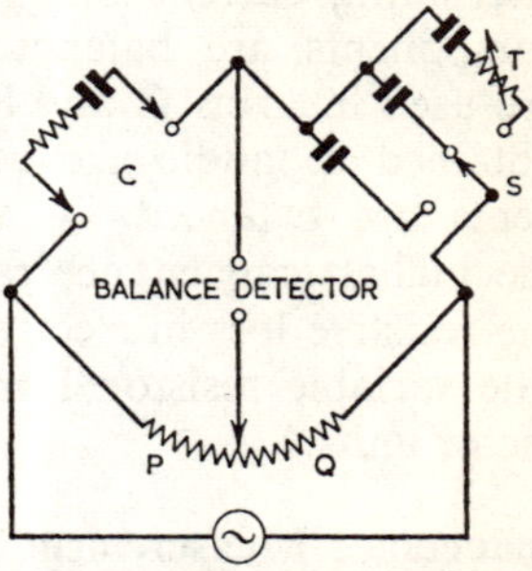

Fig. 11.—*Bridge circuit for capacitance measurement.*

resistor S by adjustment of the precision wire-wound potentiometer so that zero indication is given by the instrument I. The bridge is then said to be balanced. Under this condition,

$$R = PS/Q,$$

where P and Q are the resistances of the two sections of the potentiometer. The potentiometer may be calibrated in terms of the "standard" resistor S.

Capacitance Measurement

Fig. 11 shows one of the simplest forms of capacitance-measuring bridge based on the resistance bridge of Fig. 10. A different "standard" capacitor is switched in for each range. The balance detector must be sensitive to low alternating voltages; a high-impedance voltmeter or a magic-eye indicator is often used but headphones are very sensitive to frequencies of the order of 1 kHz. Note that the equation for balance is somewhat different from that used for resistance, since capacitance reactance equals

$$1/2\pi fC, \quad C = SQ/P.$$

A clear indication of the balance point is only obtained if the leakage and resistive loss of the capacitors are negligible. This is because a true balance can only be obtained in an alternating current bridge if both the resistive and reactive components are balanced. When good quality capacitors are used in arms C and S, a fairly clear balance is, however, obtained at audio frequencies if only the reactive components are balanced. A variable resistor T, switched in on the highest capacitance range, has to be adjusted to balance the resistive loss in electrolytic or other high value capacitors. The variable resistor T may be directly calibrated in power factor units.

Inductance Measurement

Fig. 12 shows a Maxwell bridge for the measurement of an inductance L having a resistance R.

$$L = PQC, \quad R = PQ/R1.$$

It should be noted that the inductance of iron-cored inductors depends on the frequency of the supply on which

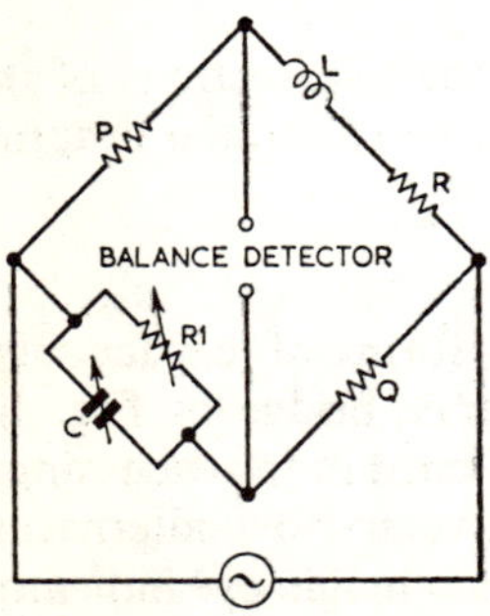

Fig. 12.—*Maxwell bridge circuit for inductance measurement.*

the component is tested and on the value of alternating and direct current carried by the component. Incremental inductance bridges are available for testing these components under specified conditions.

Self-balancing Bridges

Any form of bridge circuit can be made self-balancing by the use of a servo-mechanism, controlled by the unbalance voltage of the bridge and driving the balancing element. Fig. 13 shows the conductivity of a liquid being continuously measured. The conductivity is determined by means of a Wheatstone bridge supplied with 50 Hz a.c. If the bridge is

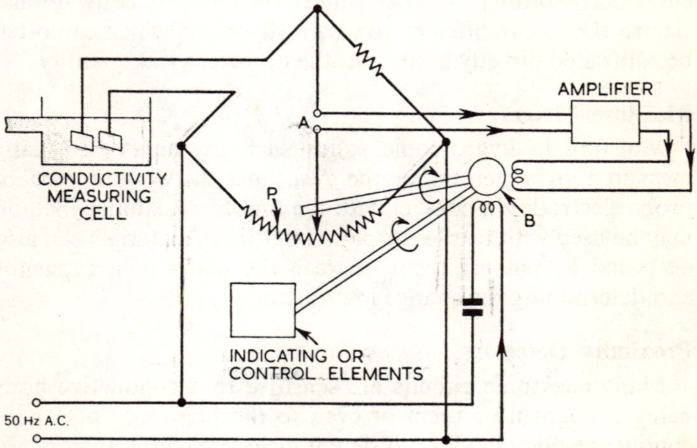

Fig. 13.—*Self-balancing bridge being used to measure liquid conductivity.*

unbalanced, an error voltage, developed between terminals A, will be amplified and used to drive the two-phase induction motor B, which is geared to the balancing potentiometer P. The error voltage will become reduced as the null or balance position is approached, hence the speed of the motor will reduce, becoming zero as the null position is reached where the error voltage falls to zero. The phase of the error voltage becomes reversed at the other side of the null position hence the motor will reverse if the system overshoots the null position. The motor may control the addition of material to the liquid so that its composition and hence its conductivity is maintained constant.

MEASUREMENT OF NON-ELECTRICAL QUANTITIES

Many non-electrical quantities may be measured by converting them into electrical quantities by means of transducers or suitable measuring heads comprising electrodes or coils. The electrical quantity is then determined by the use of conventional electrical or more often electronic instruments which can often be calibrated directly in units of the non-electrical quantity.

Moisture Meters

Moisture in hygroscopic solids such as timber is usually measured by determining the resistance between a pair of probe electrodes in contact with the timber. A similar method may be used with textiles or paper, but these materials can also be tested by passing them between the plates of a capacitor and determining the change in capacitance.

Proximity Detectors

Many electronic circuits are sensitive to a conductive body being brought near them or even to the proximity of a non-conductive body having a different dielectric constant from air. A radio frequency oscillator having an electrode connected to its tuned circuit is a typical example of such a detector. The amplitude and/or frequency of oscillation will alter as a body approaches the electrode, hence the presence of the body may be detected by an amplitude or frequency sensitive stage coupled to the oscillator and operating a relay. The electrode may be supported in a tank if it is desired to indicate when the level of a liquid or granular substance contained in the tank falls below or rises above the level of the electrode.

Other uses for proximity detectors are as burglar alarms, detecting when a moving part of a machine has reached a desired point, detecting articles on a conveyor, etc.

Dimension Testing

A proximity detector normally provides only an on–off indication of the presence or absence of a body. If a pair of electrodes is suspended in a container, the change in capacitance as the container becomes filled with liquid is proportional to the depth of the immersed electrodes. The container may form one of the electrodes. Alternatively, a float supported by the surface of the liquid can be mechanically coupled to a variable resistor.

The dimensions of solid bodies can be measured very accurately by the use of a transducer comprising a variable impedance, the value of which is changed by movement of a projecting probe. The transducer is mounted on measuring apparatus so that when the test body is brought into position, the probe is displaced in accordance with the dimension of the body.

A piezo-electric transducer carrying a stylus can be moved over a surface to determine its roughness. The amplified output voltage of the crystal provides an indication of surface roughness.

The dimensions of metal bodies can also be determined by measuring the capacitance between the surface of the body and an adjacent electrode, the opposite surface of the body being located in a datum position relative to the electrode.

Liquid Conductivity Measurement

The conductivity of pure water is extremely low, hence the concentration of a particular substance in a liquid or the proportion of impurities present may be determined by measurement of the conductance between two electrodes immersed in the liquid. It is usual to use an alternating current bridge, as in Fig. 13, to avoid polarisation at the electrodes.

Polarographs*

If the potential between a pair of electrodes immersed in a solution is gradually increased from zero, the current increases in one or more steps depending on the concentration and type of substances dissolved in the liquid. It is essential to keep the surface of the cathode clean. This is usually done by using as a cathode mercury dripping from a tube immersed in the solution: the cathode is continually replenished as the mercury drips out. The pool of mercury at the base of the container forms the anode. The current is amplified, and the graph of current against voltage plotted on a cathode-ray oscilloscope or a pen recorder. The voltage value at each step is indicative of the substance present, whilst the height of each step indicates the concentration of that substance.

pH Meters

In any aqueous solution, the product of the concentration of hydrogen ions (H^+) and hydroxyl ions (OH^-) is always the same, being equal to 10^{-14} gram molecules per litre. In acid solutions the concentration of hydrogen ions is large, hence the concentration of hydroxyl ions must be small. The reverse is the case in alkaline solutions, whilst in pure water the concentrations of the two ions are equal.

The acidity of a solution is expressed in terms of pH. This is dependent on the mathematical power of the hydrogen ion concentration.

$$\text{pH} = 1/\log_{10} C_{H^+} = -\log_{10} C_{H^+}$$

where C_{H^+} is the hydrogen ion concentration. In pure water, $C_{H^+} = 10^{-7}$ hence pH = 7. Acids have a pH value between 0 and 7 whilst in alkalis the pH value varies between 7 and 14.

Electrical methods of pH determination† usually operate by the measurement of the potential developed between a pair of

* *Journal of the British Institution of Radio Engineers*, Vol. X, May 1950, pp. 199–201.

† *Journal of the British Institution of Radio Engineers*, Vol. XIII, August 1953, pp. 401–411.

specialised electrodes immersed in the test solution. Fig. 14 shows a pair of such electrodes. The glass electrode provides a potential dependent on the pH value whilst the potential of the reference electrode (needed to complete the measuring circuit) must be independent of pH. Although thin bulbs of comparatively low resistance glass are used, the resistance of a glass electrode may be many megohms and very dependent on temperature. It is therefore necessary to measure the potential by means of a valve voltmeter having a high input impedance. The electrode potential may be expressed as

$$E = A + BT - KT\text{pH}$$

where A, B and K are constants for a particular electrode and T is the absolute temperature, i.e. degrees Centigrade $+273$. The

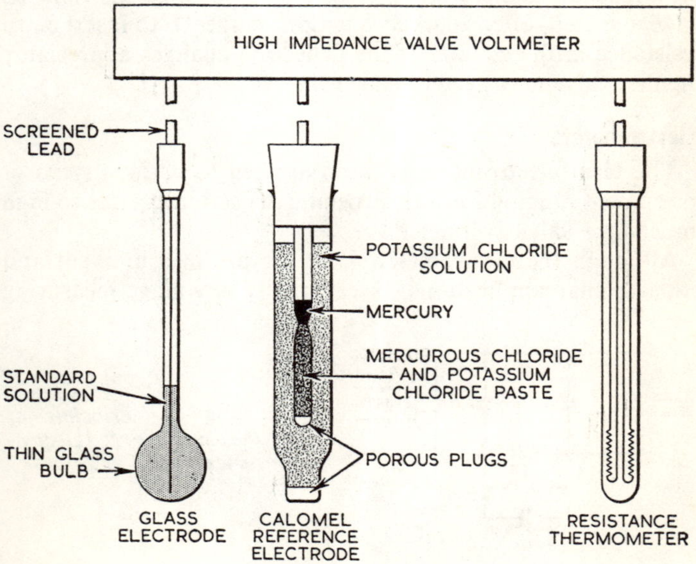

Fig. 14.—*The pH meter.*

first two terms, often known as iso-potential, are independent of the pH value of the solution tested. If the apparatus is to be directly calibrated, a compensating voltage, partly dependent on temperature, must be connected in series with the electrodes. The temperature dependent part can be derived from a potential divider comprising a resistance thermometer immersed in the test solution. Correction for temperature in the third term is achieved by means of a resistance thermometer connected in the negative feedback path of the high-impedance voltmeter amplifier.

The accuracy of a pH meter can be checked by measuring the pH value of a standard solution of known pH (buffer solution). It should be noted that it is necessary for the valve voltmeter to have a high input impedance not merely because of the high resistance of the glass electrode but because of its pronounced temperature characteristic. Some instruments have a switch, often marked g.c. (grid current), to insert extra resistance in the circuit. If the deflection changes appreciably the first valve may need replacement.

Electrometers

The term electrometer is often applied to various types of instrument ranging from electrostatic deflection devices to high impedance valve voltmeters.

Although the electrometer, used for pH measurement and other similar applications, is essentially a voltage-measuring

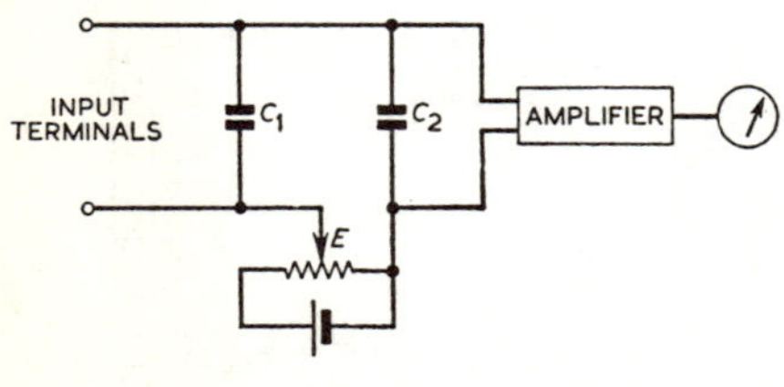

Fig. 15.—*Principle of the vibrating-reed electrometer.*

device, if a current is allowed to flow through a known resistance, the voltage developed can be used as a measure of the current. The extremely high input impedance of the electrometer permits very small currents to be measured. The minute currents developed in ionisation chambers and similar current sources are measured in this way.

A particularly sensitive type of electrometer is the vibrating-reed electrometer. The principle of this device is shown in Fig. 15. C_2 is the capacitance formed by the vibrating reed and an adjacent plate. A charge q, placed in the system, via the input terminals, when E is zero, is distributed between C_1 and C_2. The charge on C_1 is $q_1 = qC_1/C_1 + C_2$, and the charge on C_2 is $q_2 = qC_2/C_1 + C_2$.

Since C_2 is varying periodically, the charge is periodically redistributed. The amplifier amplifies the alternating voltage across C_2, and displays it on a suitable indicator. If the series voltage E is adjusted until the output from the amplifier is zero, then the voltage across C_2 is zero.

The charge supplied to C_2 by E is

$$q_2' = EC_2 \frac{C_1}{C_1 + C_2}$$

and this cancels out q_2, so that $q_2 = -q_2'$,

or
$$\frac{qC_2}{C_1 + C_2} = -EC_2 \frac{C_1}{C_1 + C_2}.$$

Then $E = -qC_1$. E is therefore a measure of the original charge q, and the result is independent of the value of C_2. In practice, E is supplied by feedback and the instrument indicates q directly.

The input resistance of this type of instrument is limited only by the resistivity of available insulators, and may be as high as 10^{15} ohms. Currents of 10^{-15} A may be measured.

CATHODE-RAY OSCILLOSCOPES

The cathode-ray tube forms the heart of the most versatile electronic instrument. Since it makes use of the deflection of

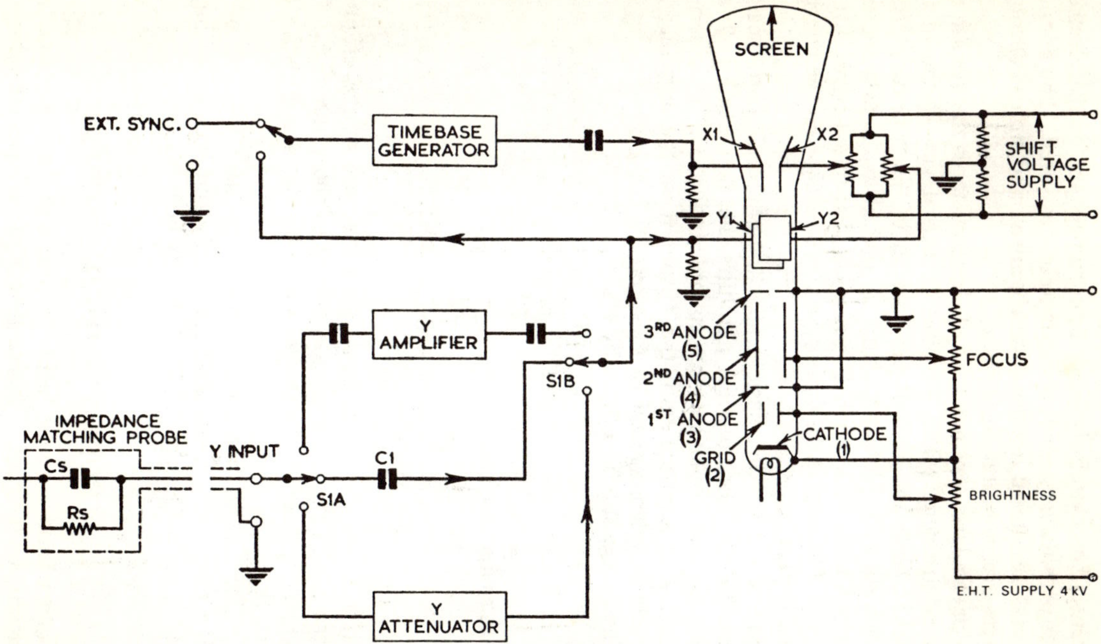

Fig. 16.—*Simplified diagram of a typical oscilloscope.*

a beam of electrons it is able to trace out the most complex waveform with almost unlimited speed.

A simplified diagram of an oscilloscope using an electrostatic cathode-ray tube is shown in Fig. 16. The tube comprises an electron gun consisting of beam-forming electrodes 1–5, deflection plates Y1, Y2, X1, X2 and a fluorescent screen. The electrons emitted by the electrically heated cathode are accelerated by three anodes and formed into a pencil-shaped beam which is focused to a point on the face of the tube. Some electrostatic tubes use only two anodes. The fluorescent screen is made by coating the face of the tube with a substance (often a zinc compound) which emits light when struck by the electron beam. The brightness is controlled by variation of the negative voltage on the cylinder (often called the grid) adjacent to the cathode. Focus is adjusted by variation of the voltage on the first anode in a two-anode tube or on the second anode in a three-anode tube.

In an oscilloscope the final anode is usually earthed and the other electrodes are connected to a voltage divider chain fed from a negative supply. The beam may be deflected upwards by applying a positive voltage to the Y1 plate or a negative voltage to the Y2 plate. Horizontal deflection may similarly be achieved by applying voltages to the X plates.

The most common use of the oscilloscope is the examination of the waveform of alternating voltages. A sawtooth-shaped voltage waveform must first be applied to the X plates to provide a timebase. The beam will thus be deflected comparatively slowly from the left-hand side of the screen to the right and returned much faster. If the timebase frequency is greater than a few cycles per second a continuous bright horizontal line will appear on the screen. When the alternating voltage to be examined is applied to the Y plates, the electron beam will trace out a graph of several cycles, then return quickly to the origin and repeat the tracing. If the timebase frequency is not an exact sub-multiple of the waveform under examination, the display will appear to move across the screen.

It is therefore necessary to synchronise the timebase generating circuit with the wave being examined.

If the timebase waveform had the same frequency as the wave being examined, the whole of one cycle could not be seen, since it takes finite time for the timebase voltage to return the beam to the left-hand side of the tube. This movement is known as flyback.

The deflection sensitivity of a cathode-ray tube is only a fraction of a millimetre per volt. It is usually necessary to amplify the test voltage before applying it to the Y plates. Deflection amplifiers are usually high gain video amplifiers having variable gain so that a useful deflection can be obtained from a wide range of input voltage. The amplification and phase shift should be the same at all frequencies over a wide band if the oscilloscope is to be used for the examination of pulse waveforms.

Y Input Circuits

In the typical oscilloscope shown in Fig. 16, a two-pole, three-way switch S1 is provided to enable the Y input terminal to be connected to the Y1 plate via a blocking capacitor C1 or via the Y deflection amplifier. An attenuator is also provided to reduce high-voltage waveforms. The switching arrangements are complicated by the need to minimise stray capacitance between the input circuit and earth. It is always necessary for each deflection plate to be connected via a d.c. path (usually a high value resistor) to the final anode so as to prevent build up of charge due to stray electrons reaching the plates. A high resistance grid leak is also necessary in the first stage of the Y amplifier.

These high impedance circuits are prone to pick up unwanted voltages by stray capacitive coupling. Screening is, however, made difficult by the need to minimise the capacitance to earth.

All the difficulties associated with the input circuit apply also to the lead connecting the input terminals to the circuit under test. If this lead is screened its capacitance may by-pass the high frequency components of the tested waveform. When testing oscillators or resonant circuits the capacitance may detune the circuit or even stop the circuit oscillating. An unscreened lead may, however, pick up unwanted voltages such as the mains voltage. When testing the input circuits of an amplifier, the lead may pick up the output signal and cause the amplifier to oscillate. Radiation from the lead to the input circuit of an amplifier may cause the same trouble when testing output circuits. To overcome these difficulties, an impedance converting probe may be desirable.

Probes

The simplest input probe consists of a small capacitor in series with the centre conductor of a screened input lead. If the value of this capacitor is made one-ninth of the combined capacitance of lead and input circuit, the loading of the oscilloscope on the test circuit will be reduced to one-tenth. The test voltage will also be reduced to a tenth of its original value, but this can be overcome by the use of the deflection amplifier. This circuit does not pass d.c. levels (whereas most deflection amplifiers do) and gives an increased loss at low frequencies depending on the resistance and capacitance of the input circuit.

An improved probe is shown in Fig. 16. A resistor R_s is connected in parallel with the series capacitor C_s. If

$$R_i/R_s = C_s/C_i,$$

where R_i and C_i are the resistance and capacitance of the input circuit, then the terminal voltage at d.c. or any frequency will be $R_i/(R_i+R_s)V_i$ where V_i is the actual voltage of the circuit tested. The input impedance will be increased by $(R_i+R_s)/R_i$ at any frequency.

Another form of probe uses a cathode follower in a small container at the end of the input lead. This gives a substantial increase in input impedance with a voltage loss of only about one-tenth, but requires an additional connection to carry its power feed.

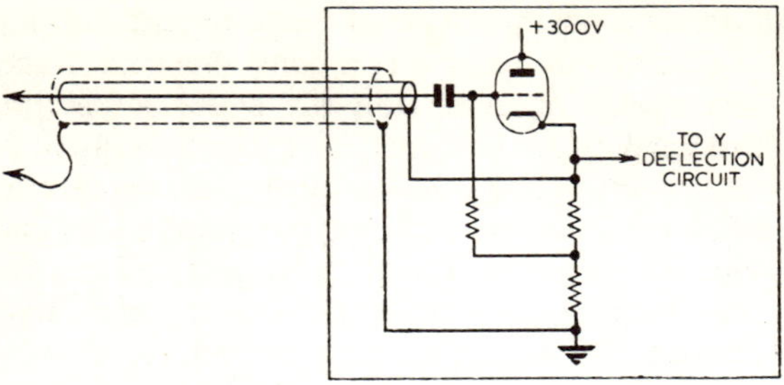

Fig. 17.—*Input circuit for matching a probe to an oscilloscope, using a double-screened input lead feeding a cathode follower.*

Another method uses a double-screened lead (Fig. 17) in which the inner screen is driven by a cathode follower contained within the oscilloscope. Since the cathode follower maintains the inner screen at approximately the same potential as the centre conductor, very little capacitive current will flow between the centre conductor and the inner screen. The current between the inner screen and the earthed outer screen will of course be provided by the cathode follower.

With modern f.e.t. input circuits, a probe may be used more as a "convenience tool" than as an impedance converter.

Double-beam Oscilloscopes

The simplest form of double-beam oscilloscope makes use of a beam-splitting cathode-ray tube having an earthed splitting plate positioned between the Y plates. A common timebase voltage fed to the X plates results in two horizontal traces appearing on the screen, the vertical position of the traces being dependent on the setting of the two shift controls feeding direct voltages to the two Y plates. A voltage waveform fed to the Y1 plate will be displayed on one trace and another waveform fed to the Y2 plate on the other. Interaction between the two beams is usually negligible except when very high frequencies or pulses having very short rise times are examined.

Two or more traces may be displayed on a standard cathode-ray tube if an electronic switch is used to supply a square wave or staircase waveform to the Y plates. The frequency of the switching wave must be a sub-multiple of the frequency of the waves being examined. A gating circuit is also necessary to ensure that only the required input waveform is displayed on each trace.

FREQUENCY MEASUREMENT

Frequencies are usually measured by comparing them with a known standard frequency. This standard frequency may be provided by a crystal oscillator, a tuning-fork, or a precision RC oscillator, depending on the degree of accuracy required. By applying the unknown frequency to the Y plates of an oscilloscope and the standard frequency (or a suitable multiple or sub-multiple) to the X plates, Lissajous figures will be obtained. These will become stationary when an exact relationship exists between the two frequencies. Thus if the two frequencies are nearly equal an ellipse will be obtained, oscillating between two limiting diagonals. When the frequencies are exactly equal the ellipse becomes stationary.

To measure frequencies at intermediate points, a subsidiary oscillator is required which can be set or calibrated at cardinal points from a frequency standard. This is then used to provide the reference frequency, being varied until it comes into coincidence with the unknown one.

An alternative method is to count the number of waves in a given period, say one second. For this purpose decade counters may be used. These consist of a series of units each of which records the arrival of pulses up to ten, when it then passes on a pulse to the next unit and recommences its own count. Various forms of counter tube give numerical representation of the number of pulses received. A bank of four such tubes will count up to 10,000 pulses, and by arranging to "gate" the device so that it remains on for exactly one second any frequency up to 10 kHz can be measured (to an accuracy of

1 Hz). The gating impulse can be derived from a crystal standard suitably subdivided.

TRANSDUCERS

A transducer may be defined as a power-transforming device for insertion between the electrical, mechanical, or acoustic parts of a system of communication.

Measurement of Displacement

Mechanical displacements in the range of 0·0001 to 0·01 in. may be converted to electrical signals by a linear differential transformer. This consists (see Fig. 18) of three coils wound on a common former. The centre coil is energised from an alternating current source, while the outer pair of coils are connected in a series-opposing configuration. With the movable ferrite core centrally positioned, signals induced in the outer

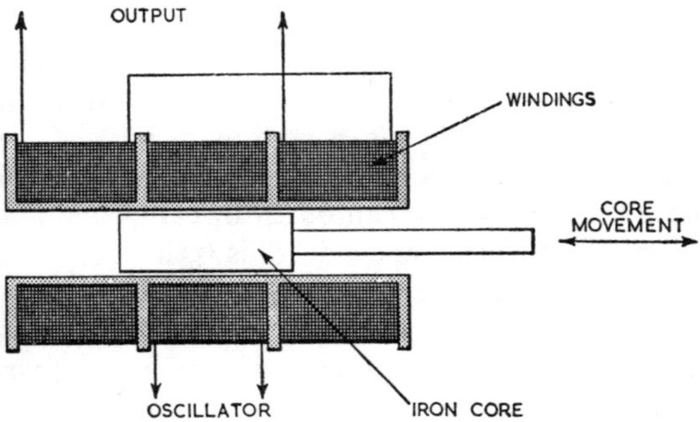

Fig. 18.—*The differential transformer.*

coils cancel and the resultant electrical output is zero. Movement of the core away from centre induces a larger signal in one of the outer coils than the other and gives rise to an output which can be made proportional to core movement. Equal core movement either side of the zero position will give equal electrical output, differing only in phase, and to determine the true position of the core a phase-sensitive detector is essential. The relationship between core movement and electrical output is only linear for small movements of the core around the balance condition.

Measurement of Proximity

The change of electrical capacitance between two metallic plates is commonly used to determine proximity. While the actual capacitance change may be small and difficult to measure, the effect of small capacitance changes on "hi-Q" tuned

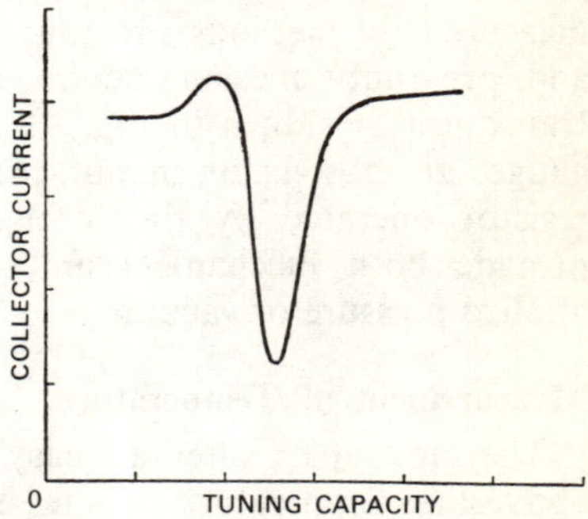

Fig. 19.—*Collector current/tuning capacitance relationship for a crystal oscillator.*

circuits is quite marked. A crystal oscillator, for example, tuned just off resonance, will draw collector current which is nearly

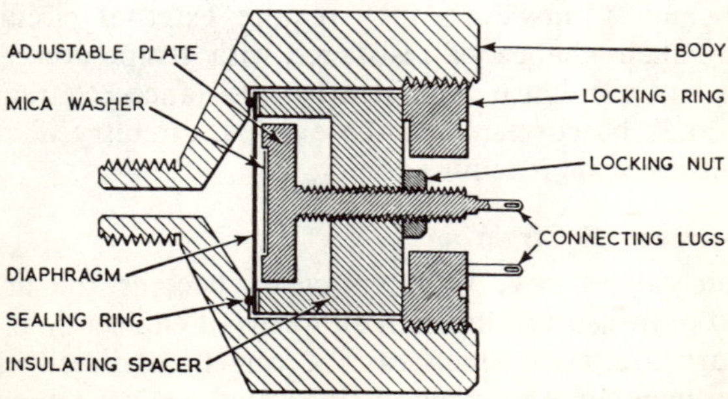

Fig. 20.—***Cross-sectional view of a simple capacitance transducer.***

proportional to small changes of capacitance shunted across the tuned circuit (see Fig. 19). Circuits of this type can form the basis of sensitive and stable proximity detectors. Ferrous and non-ferrous material in the form of a core or diaphragm may be made to approach the inductive arm of tuned circuits with similar results to those stated above. A cross-sectional view of a capacitance transducer is shown in Fig. 20.

Measurement of Pressure or Vacuum

Transducers for these purposes normally first convert the pressure or vacuum to a mechanical movement which is then measured by methods previously described for displacement and proximity measurement. A simple capacitive pressure transducer is shown in Fig. 20. Bourdon tube type pressure gauges are now being manufactured which include a variable resistor operated by the mechanical linkage. Such gauges provide both mechanical and electrical indications of the applied pressure or vacuum.

Measurement of Temperature

Thermocouples offer an easy and inexpensive method of converting temperature changes into electrical signals. Various combinations of metals are used, the noble metals and their alloys being preferred for high temperatures or arduous service. In the range 0–100°C thermistors are more sensitive than thermocouples; however, they require external circuitry to measure their change of resistance with temperature. They possess a non-linear temperature/resistance characteristic which must be corrected in the external circuitry if a linear temperature scale is required.

Measurement of Strain

Strain gauges have in recent years become the accepted method of measuring strain in all forms of engineering. Many types are available which can be cemented directly to the stressed member, and when compensated against temperature changes will give readings of strain under the most arduous conditions. Bonded gauges, that is those which may be fixed directly to the stressed area, are formed by winding a fine resistance wire on a strip of thin card or paper, or by electrically depositing a zig-zag pattern of metal on a thin insulated backing. The gauge is cemented with a resin glue to the stressed surface in such a manner that the forces tend to stretch the thin wires or foil. This in turn causes the gauge resistance to increase and the changes are detected by a Wheatstone bridge circuit. Subsequent amplification may be employed to increase the range of measurement. The bridge circuit may be fed with

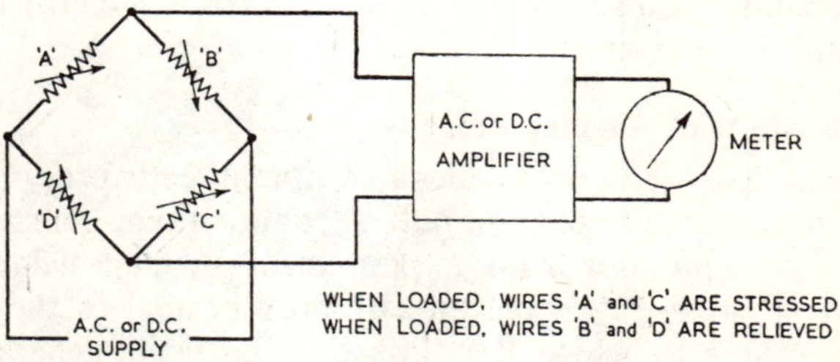

Fig. 21.—*Circuit of an unbonded strain gauge.*

direct or alternating current. Due to the small changes of resistance that take place, gauges must be compensated against changes of temperature. The method usually employed is that of lightly attaching a second gauge to the area under strain, the electrical connections of this being brought out to form an opposing arm of the Wheatstone bridge. While both gauges will be affected by the temperature changes, only the gauge which is firmly attached to the stressed area will produce a resistance change when stressed. Unbonded gauges consist of fine resistance wires stretched between insulated supports. Alternate wires are tensioned and slackened as pressure is applied to one of the supports. The wires form the arms of a bridge circuit, which when energised gives an output signal proportional to the applied pressure. The assembly is normally enclosed in a small metal case, and the mechanical movement is applied through a flexible diaphragm at one end. A typical circuit for an unbonded gauge is shown in Fig. 21.

Semiconductor Strain Gauges

The piezoresistive characteristics of semiconductors have been utilised in the development of strain gauges which are many times more sensitive than conventional wire strain gauges. A silicon strain gauge may exhibit as much as a 5 per cent change in resistance value for less than 0·05 per cent change in length. The temperature sensitivity of such gauges requires compensation, and this is usually achieved by the use of two similar gauges, one of which is used as a thermistor to measure the temperature.

Measurement of Angular Velocity

The tacho-generator provides an output directly proportional to the rate of change of the input. A common type consists of a small d.c. generator with a permanent magnet field. The output of this is a voltage directly proportional to the speed of the motor to which it is linked. The tacho-generator is the basic sensing device used in many feedback control systems.

Hall-Effect Devices

See Section 3.

Further Electronic Measurements

The range of electronic measurements is immense and only a few basic techniques have been given here. The reader is referred to other texts for more complete data on measurement techniques in the laboratory in order to ensure accuracy* and on the basic processes of more complex instruments.†

* Scroggie, M. G., *Radio and Electronic Laboratory Handbook*, 8th edition, Newnes–Butterworths, 1971.

† Edwards, D. F. A., *Electronic Measurement Techniques*, Newnes–Butterworths, 1971.

14

ELECTRONIC CONTROL

Many of the processes of electronic control circuitry depend for part of their action on the switching and gating of transistor circuits. The reader should therefore familiarise himself with the information on this subject given in Section 8.

A very large number of electronic control systems follow a set pattern. A physical quantity, for example, movement, pressure, flow, light intensity or radioactivity, must first be converted into an electrical signal. This signal, which may require amplification, is then used to operate a power system which produces the desired control operation. The tacho-generator when coupled to an electric motor will produce an electric signal related to the speed of the motor, the photocell will provide an indication of light values, the microphone a signal proportional to sound intensity and various other forms of transducers signals proportional to the physical quantity under control.

In contrast with audio and radio applications, where active devices are called upon to amplify complex waveforms of widely varying frequency, transistors and valves used for control purposes are usually required to handle a signal which may be slowly varying, aperiodic or requiring d.c. amplification. D.C. amplification is common in control systems in order to present the control information at a satisfactory level to operate the power conversion stage, which may, for example, be a grid-controlled thyratron or ignitron (this is called a static arrangement) or the field of a rotating machine.

Ideally, d.c. amplifiers should be free of drift. If the phase lags in the system exceed 180 degrees and the gain is greater than unity the circuit will oscillate. In industrial control applications this oscillation is generally of relatively low frequency, and is called "hunting". It may be corrected by incorporating in the system feedback circuits or phase-shifting arrangements, see Fig. 1.

It is usually necessary to start by establishing a datum or reference level. The signal from the transducer is compared with this and the resultant error amplified and used to provide

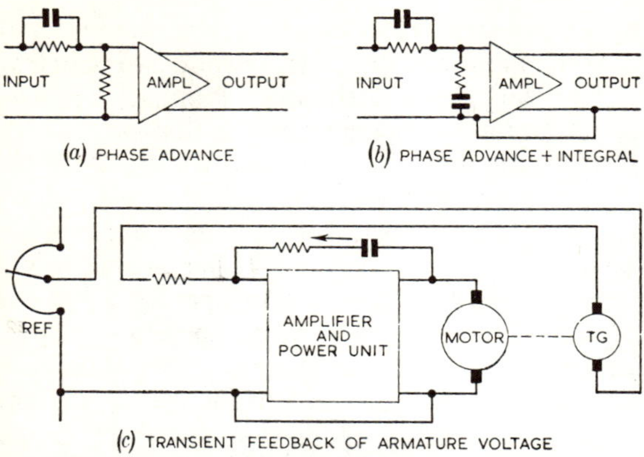

Fig. 1.—*Use of phase-shifting components and feedback to reduce hunting. TG in* (c) *is a tacho-generator.*

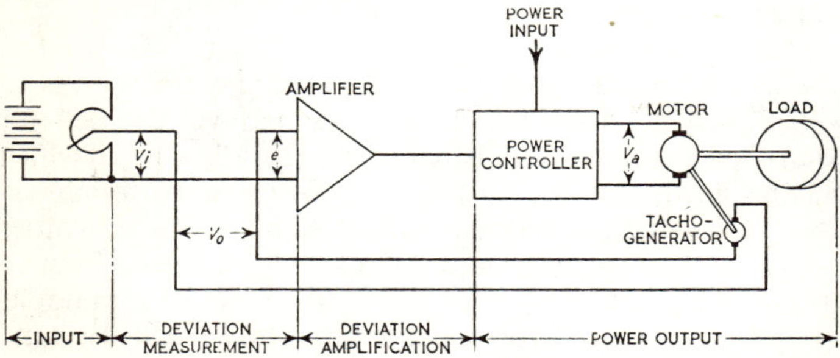

Fig. 2.—*Closed-loop electronic feedback control system.*

the control required. The reference level may be provided by a neon reference tube, zener diode or stabilised power supply.

A basic closed-loop (i.e. no operator is required to "close" the control loop) electronic feedback control system is illustrated in Fig. 2. In this example the speed of the motor is continuously measured by the tacho-generator, the output voltage of which, V_o, is proportional to the speed of the motor. This voltage is compared with an appropriate reference voltage,

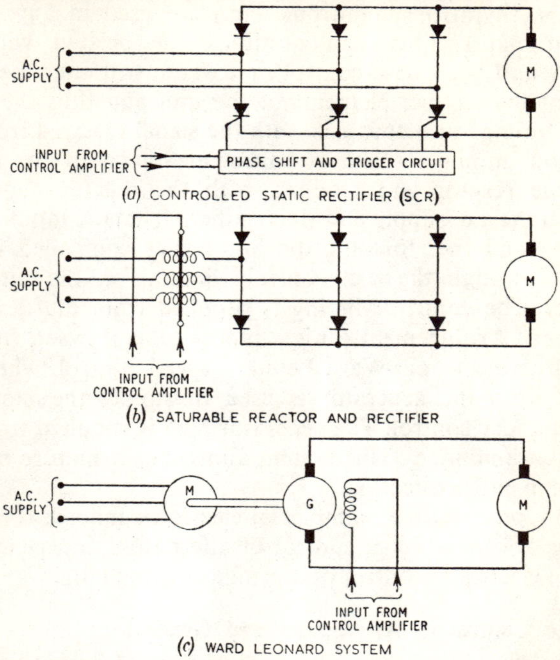

Fig. 3.—*Three examples of power control arrangements* (*using a d.c. motor in each case*).

V_i, which in this example is taken from a potentiometer supplied by a battery. The deviation e between V_o and V_i, the error signal, is measured, amplified by the deviation amplifier (often referred to as a servo or control amplifier) and then used to control the power controller which controls the input power to the motor.

The power controller is connected between the source of power (e.g. the mains supply) and the motor to be controlled, and provides a flow of d.c. power to the motor adjusted according to the error signal from the deviation amplifier. Three examples of a power control arrangement are shown in Fig. 3. In (a) controlled static rectifiers (e.g. silicon-controlled rectifiers, thyratrons, ignitrons, etc.) are used in any of the conventional rectifier configurations. Associated with the rectifier unit is a phase-control unit which will alter the firing angle of the rectifier elements of the unit and thus the mean output voltage in accordance with the signal received from the deviation amplifier. The arrangement shown in (b) uses a saturable reactor and rectifier, with the reactor connected between the a.c. supply and the rectifier. The a.c. input to the rectifier (and therefore the d.c. output) is controlled by the current through the d.c. control winding on the saturable reactor. The control winding is supplied from the deviation amplifier. Arrangement (c) is a motor-generator set, forming part of an electronic Ward Leonard speed control, where the output from the generator is used to supply the motor or motors under control. The generator field is supplied from the deviation amplifier. This system allows the armature current to flow in either direction.

Two representative examples of electronic industrial control systems are the voltage control of alternators and generators and the electronic control of a profile copying lathe.

Voltage Control of Alternators and Generators

In many industrial processes a highly stable supply voltage is necessary. The voltage is required to be kept constant at the selected value regardless of the load on the system, and must be restored to the set value in a very short time after the occurrence of transient demands. The technique used follows the usual closed-loop, error-actuated servo pattern. In the case of an alternator, the a.c. supply is first rectified and the resultant d.c., after filtering, is compared with a highly stable source, which is usually provided by a gas-filled reference tube (these are capable of providing a long-term accuracy of the order of $\pm 0{\cdot}1$ per cent). If the supply voltage varies from that of the reference source an error signal is produced which, after d.c. amplification, is used to control several parallel-connected valves placed, as shown in Fig. 4, in series with the machine

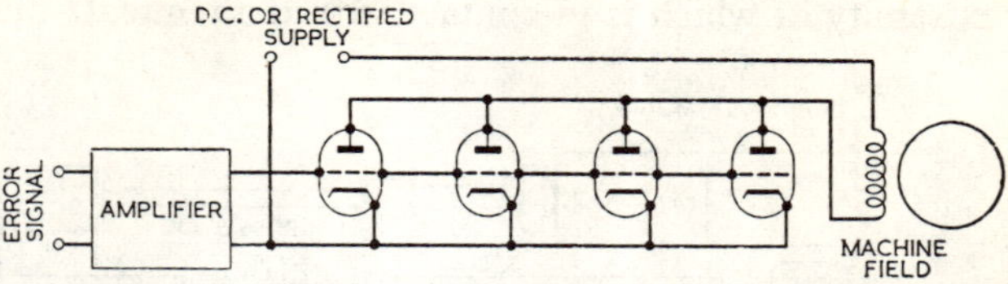

Fig. 4.—*Use of parallel-connected valves in series with the supply to the field windings of the machine to control the machine field voltage.*

field. The d.c. supply for the field is produced externally, the valves in series with it being used solely as control valves, not as rectifiers. It is necessary to have several valves in parallel in order to handle the field currents involved.

The first stage of the amplifier must be designed to have little input drift as this factor, coupled with the accuracy of the reference element, determines the optimum performance of the system. The use of valves in series with the machine field enables extremely rapid adjustment of the machine field voltage to be achieved.

Electronic Control and Numerical Control

The copying lathe is widely used for automatically producing an accurately machined replica of a master, which may be in the form of a template or pattern. It is common today for them to be electronically controlled. The layout of a typical machine together with associated electronic control circuits (in block form) are shown in Fig. 5. The tracer head is rigidly coupled to the cutting tool via the saddle. The saddle and cross-slide motors can run in either direction, independently or together, and make the cutting tool follow the outline of the template in accordance with signals provided by the electronic tracer head. For the sake of clarity a simple template is shown: in practice any shape, re-entrant or otherwise, can be accurately reproduced.

The tracer head consists of four electro-magnets symmetrically disposed around a ferromagnetic rod, termed the stylus, the extremity of which is in contact with the template during

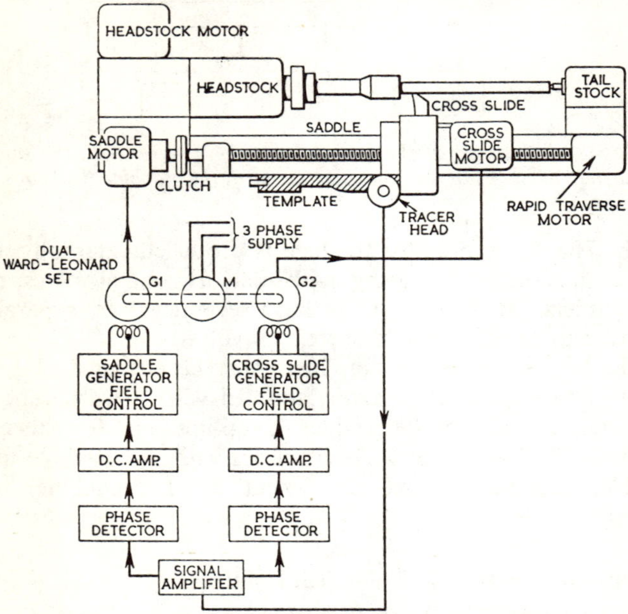

Fig. 5.—*Control system for a profile copying lathe.*

the tracing operation. The stylus is supported at one end only, and can thus be deflected from its neutral position. A pick-off coil is attached to the stylus rod, and the four electro-magnets are fed with a.c. in such a manner as to produce a rotating field. Movement of the pick-off coil within the magnetic field induces in it a voltage whose magnitude is related to the amount of the deviation and whose phase is dependent on the direction of displacement. The signal from the head is amplified and passed through a phase-sensitive detector, ultimately appearing in a form suitable for controlling the power valves which control the field of the generators supplying the saddle and cross-slide motors.

The control chain for each motor is very similar, and it is therefore necessary here only to describe one generator-motor

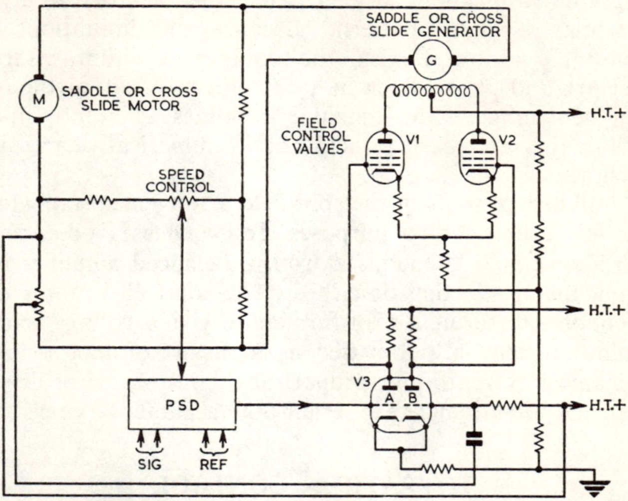

Fig. 6.—*Simplified circuit of a split-field generator control system. **PSD** is the phase-sensitive detector.*

combination. A simplified circuit of this is shown in Fig. 6. The output from the phase-sensitive detector is applied to a balanced twin triode differential amplifier (V3) directly coupled to the two pentodes which control the current to each half of the split-field generator in the dual Ward Leonard set. Under steady-state conditions the input to both halves of V3 will be identical. If, however, an error signal produced by the deflection of the stylus is applied to the circuit the "error" will be amplified by the differential amplifier which will apply different bias voltages, depending on the magnitude of the error signal, to the pentode control valves. These will then feed different currents to the two halves of the split-field winding. Thus the generator output will vary in magnitude and polarity according to the amplitude and sense of the error signal, causing the associated motor to run at varying speeds and in one direction or the other.

The signal provided by the tracer head can be replaced by a programmed magnetic tape or paper tape recorded to give a particular working pattern. Because the limitations on positioning no longer exist—the tool may be withdrawn from the work and placed back in a different position, or the tool may be changed—the machine becomes inherently more flexible: this is the basic principle of the numerically controlled machine.

Simplified versions of the split-field arrangement are widely used for various control purposes. In many cases a d.c. motor with a split field is energised from a balanced amplifier in a similar manner to that described. The controlled motor may be coupled to toroidal transformers to give a voltage-control system. It may also be used as a means of motorising a mechanical valve, to give proportional control, or for driving balancing potentiometers in certain high-accuracy servo systems.

THYRATRON CONTROL

The thyratron, which finds wide application in industrial electronic control, is in effect an electronic switch: the circuit is closed by means of the control grid and opened by interrupting the anode current. As is pointed out in the section on Thermionic Devices (Section 2), the grid exercises no control at all over the flow of anode current once the thyratron conducts ("strikes"). The thyratron will therefore not amplify complex waveforms or continuously varying voltages. The usual method of controlling a thyratron is to reduce the negative bias on the grid to the point at which the valve fires: this voltage, the critical grid voltage, is related to the value of the instantaneous anode potential. The anode is frequently supplied with a sinusoidal a.c. supply and, as a result, the critical grid voltage will vary accordingly.

When two thyratrons are connected in inverse-parallel and allowed to run uncontrolled, they will pass a complete cycle of a.c. If, on the other hand, the tubes are prevented from firing by means of grid control until the anode potential has reached its maximum or beyond, then "chopped" sine waves are reproduced, their average value, and hence the power applied to the load, being correspondingly lower. This type of circuit is widely used for the control of low power resistance welding plant, fluorescent-lamp dimming and high-speed electronic switching.

A commonly used arrangement is the operation of two or more thyratrons in the usual rectifier circuits providing a continuously variable d.c output by means of grid control. This is the basis of a large number of electronic control schemes

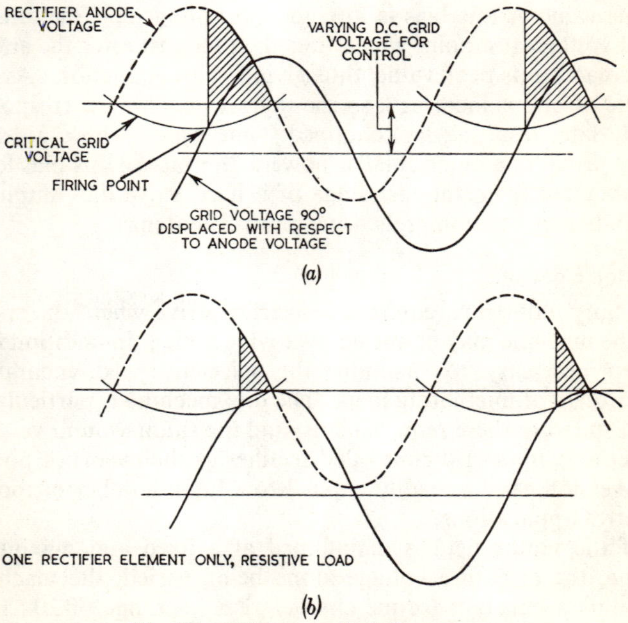

Fig. 7.—*Grid control of thyratrons.*

embracing electronic motor control, voltage control for alternators, coiling and reeling drives, and electronic Ward Leonard systems.

There are various means by which control may be accomplished and these include a simple d.c. bias arrangement, also a system utilising a "peaky" waveform suitably phase shifted. A widely used arrangement which is readily adaptable to closed-loop servo systems is that which incorporates a phase shifted a.c. waveform of fixed amplitude superimposed on an adjustable negative bias. This system is indicated in Fig. 7 and for simplicity only one rectifier element operating into a resistive load is being considered.

The grid voltage is 90° displaced with respect to the voltage at the thyratron anode, and is superimposed upon a negative bias supply which can be varied. It can be seen from Fig. 7 that if the value of this bias is low, the a.c. voltage cuts the critical grid voltage at a point which fires the tube just after the anode has reached its peak value, thus giving half-conduction. As the value of bias is increased, so the thyratron is fired correspondingly late, resulting in "chopped" sine waves whose average level provides a much smaller power. In practice this bias level is obtained from the last stage of a hard valve d.c. amplifier included in the comprehensive control system.

Motor Control

Many industries require an electrical drive where the speed of the machine may be varied over wide limits. In addition it is often necessary to maintain the selected speed accurately regardless of fluctuating load. The d.c. machine is particularly well suited to these requirements, and the shunt-wound version together with a grid-controlled rectifier as the means of power conversion can be readily applied to a large number of motor control applications.

If the motor field is maintained at a fixed and maximum value, the armature voltage alone being varied, the machine exhibits a constant torque characteristic, a range of 20 : 1 or more being readily attainable. If the field is also supplied from thyratron grid-controlled rectifiers a further 3 : 1 variation is possible, providing an overall range of 60 : 1.

The essentials of a typical electronic variable-speed drive system are indicated in Fig. 8, the mode of operation being as follows. The armature voltage, or alternatively a tacho-generator, is used as an indication of motor speed, and is compared in a subtractive sense with a highly stable reference potential. Any deviation of the motor or tacho-generator voltage from the level set by the reference causes a difference or error signal to be produced, with a polarity dependent upon the sense of the deviation and a magnitude proportional to the amount of the deviation. This error signal is fed to the d.c.

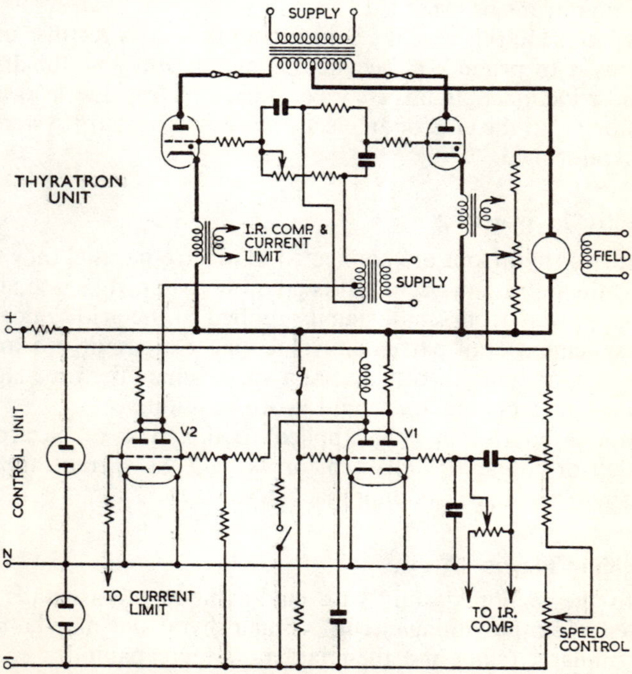

Fig. 8.—*Simplified circuit of an electronic variable-speed drive unit. Note phase-shift network in the grid circuit of the thyratrons.*

amplifier V1 and V2 which includes stabilising circuits as a precaution against hunting. The varying potential across the anode load resistance of V2 provides the necessary bias to control the thyratrons in the manner previously described.

In order to compensate for the poor load/speed characteristic which would result from a control of armature voltage alone, a further voltage proportional to the motor current, and hence its load, is injected into the control system. This is termed IR drop compensation and is achieved by small current transformers situated in the thyratrons' cathode circuits. These current transformers are also used to provide an over-riding signal should the maximum current which can be accepted by the thyratrons be exceeded.

Whereas in this case the control and thyratron rectifier units are used to produce a varying d.c. supply for a motor drive, almost identical circuits may be utilised to feed the field of a motor or, in the case of an electronic Ward Leonard system, a generator field.

Electronic Switching

If two thyratrons are connected in inverse-parallel they will pass the full a.c. cycle, and it is possible to control reasonably large powers from small signals applied to the grid. Such an arrangement is of particular value where a circuit has to be switched very frequently and at a speed where the time lag of a mechanical contactor would prove prohibitive.

Power control may be applied to this type of electronic switch or contactor and used for varying the current applied to small resistance welding machines.

Welding Timers

In the section dealing with cold-cathode tubes details are given of simple timing circuits. Small thyratrons may be used in similar circuits and their larger power capabilities enable contactor coils to be placed directly in series with the anode supply.

For simple applications, only the welding period itself has to be determined whereas, on other occasions, circumstances may demand a four-interval timer providing squeeze, weld, forge and re-set times.

Synchronous timing is used in conjunction with power control of the ignitron contactor, and thyratrons are used for providing the firing impulse. In the seam welding process, which necessitates heat and cool periods alternately, care must be taken to ensure that if one timing period ends with a positive half-cycle, then the succeeding weld must commence with a negative half-cycle. This avoids saturation of the welding transformer and thyratrons used in a lead/trail circuit are commonly employed for this purpose.

Installation

Mercury vapour thyratrons must be mounted vertically, base downwards. On the other hand, gas-filled types may be mounted in any position between the horizontal and base downwards.

Thyratrons are more sensitive to variations in filament voltage than high-vacuum tubes and care must be taken to ensure that the manufacturers' limits are not exceeded. The filament current requirements of industrial type thyratrons are considerable so that voltage measurements should be made directly at the base of the tube.

In certain industrial applications, thyratrons are required to operate in the vicinity of high frequency dielectric loss or induction heaters. In these circumstances they must be suitably shielded from strong r.f. fields otherwise random firing due to ionisation of the gas is liable to occur.

Since small variations in the firing point normally occur, thyratrons are able to radiate electrical noise. If sensitive communication equipment is used in the vicinity the interference can be troublesome. This condition can be cured by means of suitable suppression devices which must be situated as close to the thyratron anode caps as possible.

THE SILICON-CONTROLLED RECTIFIER

The silicon-controlled rectifier, the operation of which is described in Section 3, is a semiconductor device which behaves in a manner similar to the thyratron or grid-controlled rectifier. It has, however, the advantages of: (1) very low forward voltage drop in the conducting state (usually less than 1 V) resulting in high efficiency; (2) fast trigger and recovery times—only a few microseconds; (3) absence of heater; (4) high temperature operation. Applications include equipments that have previously used thyratrons and grid-controlled rectifiers. It may also be used in some applications instead of the magnetic amplifier or power transistor. Examples of its use include static switching, d.c. motor control, d.c. power regulation, dynamic braking, and welding control.

Once the silicon rectifier is "fired" by a gate signal, the gate has no further control, and conduction ceases on interruption of the anode current. The gate signal required is of the order of 2–4 V, the current required being between 15 mA and 400 mA depending on the rating and type of s.c.r. The maximum average gate power permitted is approximately 500 mW, with a limiting peak value of 5 W. Firing circuits can be used which give (per cycle) a single pulse output, a rectangular pulse or a train of pulses, e.g. from a blocking oscillator or multivibrator. Control over the firing circuit may be effected by a direct voltage or current signal from the control amplifier.

In many control applications the firing point of the s.c.r. is varied within the positive half-cycle by using a firing circuit providing pulses whose phase can be altered with respect to the a.c. supply. For accurate control the firing pulse should be steep fronted: the spread in characteristics and temperature changes affecting s.c.r.s make it difficult to achieve accuracy with sinusoidal trigger circuits such as are often used with thyratrons, where the output from an a.c. winding on the transformer is passed through a variable phase delay, although this system can be used with s.c.r.s where only rough control is required.

Integrated circuits are also available that compare the a.c. phasing with the input gating signal. The comparison determines that the s.c.r. switches only when the a.c. passes through zero volts. This leads to the name *zero-voltage switch.* The fact that no current is being switched at this instant eliminates any possible interference effects that might otherwise result from using s.c.r.s.

Fig. 9 shows an equipment using s.c.r.s in a three-phase full-wave circuit. As the inverse voltage rating of earlier s.c.r.s was smaller per unit than that of thyratrons, it was necessary to use several in series per arm when the equipment was to be used from a standard 415 V supply. One half of the bridge consists of semiconductor diodes (not controlled) and the other half of s.c.r.s. As there are many types of firing circuit which can be used to control s.c.r.s, this part of the circuit is not shown in detail; it is assumed that a variable-voltage input to

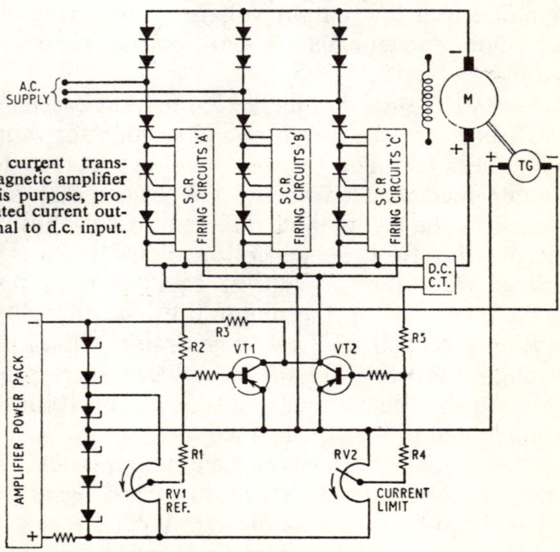

D.C.C.T.–D.C. current transformer. A magnetic amplifier is used for this purpose, providing an isolated current output proportional to d.c. input.

Fig. 9.—***Motor control system using silicon-controlled rectifiers.***

the firing circuits will provide the necessary control of phase shift. The variable voltage is supplied by the control amplifier, which uses transistors and has provision for current limit. Tacho-generator feedback is shown.

Speed Control.—The speed reference voltage is obtained from RV1, connected across a zener diode chain supplied from a d.c. source. This reference is compared with the voltage (proportional to speed) from the tacho-generator driven by the motor. The comparison is made in resistors R1 and R2, and the deviation signal is applied to the base of transistor VT1. Transistors VT1 and VT2 have a common load resistor R3, and as the base of VT2 is made positive by the voltage from RV2 through the resistor R4, transistor VT2 is cut off. When the slider of RV1 is at the minimum speed position (i.e. slightly negative to the common line of the amplifier and transistor emitters), transistor VT1 is in a conducting state. Its collector potential is slightly negative to the common, resulting in a low output voltage to the firing circuits. This condition corresponds to zero output from the armature rectifier.

As RV1 is turned to increase speed, the base/emitter voltage of VT1 becomes less negative and its collector voltage becomes more negative The (negative) voltage output to the firing circuits increases, advancing the firing point of the s.c.r.s, increasing the d.c. output to the motor and thus increasing its speed. When the tacho-generator voltage balances the reference voltage from RV1 a state of equilibrium is reached. Any increase of load on the motor will tend to reduce its speed, causing a reduction of tacho-generator voltage. The current through transistor VT1 will be reduced, increasing the output signal to the firing circuits and the output from the armature rectifier, and thus restoring the motor speed.

Current Limit.—This is effected by transistor VT2, normally held non-conducting by its positive base signal received from the current-limit potentiometer RV2. When the voltage signal proportional to current, from the d.c. current-transformer in the supply line to the motor, exceeds the voltage from RV2, transistor VT2 begins to conduct. As it has the same load resistor as VT1, R3, the collector voltage is reduced; this retards the firing point of the s.c.r.s and reduces the output voltage to the motor, preventing any further rise in motor current.

ELECTRONIC WARD LEONARD SYSTEMS

For high power application and also where certain special features are required, the electronic Ward Leonard set as the means of power conversion offers a number of inherent advantages. In these circumstances the generator field is supplied from thyratron or silicon-controlled rectifiers, the general arrangement being similar to that described in the

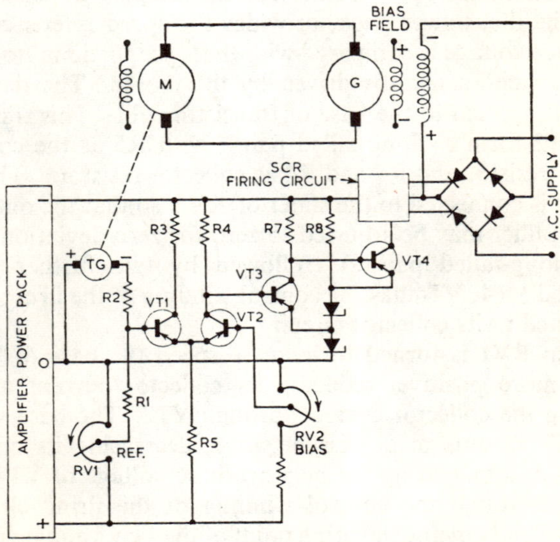

Fig. 10.—*Simplified circuit of an electronic Ward Leonard motor control system.*

foregoing section. This arrangement allows for regenerative braking on calling for a lower speed (not easily accomplished with static rectifiers) and the stored energy together with the thermal capacity of the power converter permits the application of high peak powers to the controlled motor.

A simplified circuit of a Ward Leonard speed control equipment is shown in Fig. 10. The field of the generator is supplied by a single-phase controlled rectifier. The type of firing circuit used (but not shown in detail) is that incorporating a saturable reactor and requiring a current output from the control amplifier. The control winding of the saturable reactor is in series with the output of the last stage, VT4. The generator has a bias field in addition to the main control field; this is excited with the opposite polarity to the main field, enabling the regulator to reduce the generator output to zero when necessary.

Potentiometer RV1, connected across part of the output from the d.c. power pack, provides the speed reference. The reference voltage is compared with that (proportional to speed) from a tacho-generator driven by the motor. The deviation signal is applied to the base of transistor VT1. This transistor and VT2 form a "long-tailed pair", with R5 as the common emitter resistor and R3 and R4 as collector resistors. The base of VT2 is connected to the slider of RV2, so that the output of the amplifier may be adjusted to zero for zero deviation input. The "long-tailed pair" is followed by two further stages, VT3 and VT4; VT4 has the control winding of the firing circuit connected in its collector circuit.

When RV1 is turned to increase speed the base of VT1 is made more positive, reducing its collector current and increasing the collector current through VT2. The base voltage of VT3 is thus made less negative, decreasing its collector current and increasing the negative base voltage of VT4. The current through the control winding of the firing circuit is increased, advancing the firing point of the s.c.r.s and increasing the field excitation to the generator. The generator voltage and thus motor speed increase, raising the output from the tacho-generator to balance the increase in reference from RV1.

THE IGNITRON

This is a single anode, water cooled, mercury pool tube used mainly for the control of circuits where high peak-powers of relatively short duration are encountered. It is thus especially useful in the control of resistance welding plant, where it can be used to replace heavy duty contactors. An important advantage of this type of control lies in the fact that it is an almost inertia-less system closing the circuit in one hundredth of the time taken by a mechanical contactor. The conduction period may be altered by retarding the point of ignition, which provides a means of varying the power of the system from small input signals.

The structural arrangement of a typical welding ignitron is shown in Fig. 11. The anode is made of graphite, the ignitor of refractory material, usually boron carbide, and the whole structure is contained within a stainless steel water jacket.

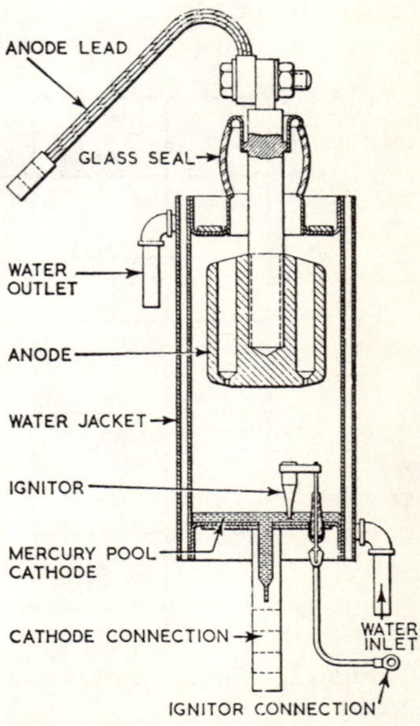

Fig. 11.—*Internal structure of a welder ignitron.*

The rectifier type of ignitron is used for conventional rectification purposes and differs slightly from its welding counterpart. It is often required to operate at higher peak inverse voltages, and anode/cathode baffles are interposed to obviate the possibility of arc-back. A spare ignitor is included together with auxiliary anodes so that the cathode spot may be maintained should the current fall to a very low value as may well happen if the load produces a counter e.m.f.

Ratings

The welder ignitron is designed to deal with high peak currents occurring intermittently. The choice of tube for a given application is thus related to the duty cycle of the welding machine. Examination of the rating curves given in Fig. 12 enables the maximum primary current available to be established. Due to the low thermal capacity of electronic tubes in general, the duty cycle must be related to a maximum period, termed the averaging time.

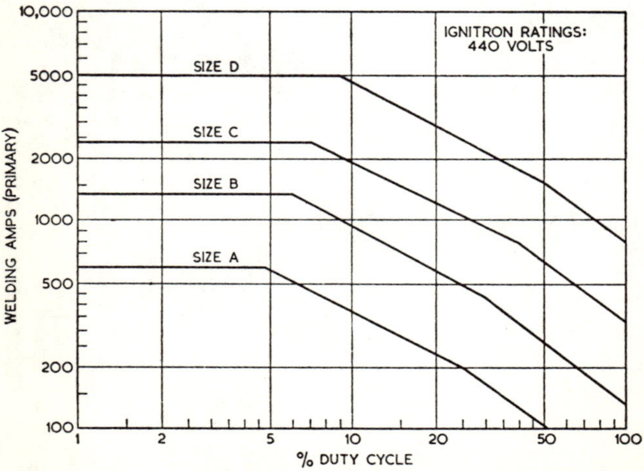

Fig. 12.—*Rating curves for welder ignitrons (international sizes).*

Welding Control

When used as an electronic contactor two ignitrons should be connected in inverse-parallel in order that the full a.c. cycle may be reproduced. In this case no power control (other than transformer taps) is used. The power for ignition purposes is derived from the mains supply, a rectifier serving to prevent reverse current, which would damage the ignitor. The ignitron contactor depicted in Fig. 13 is normally used in conjunction with a simple electronic timer incorporating thyratrons, high vacuum or cold-cathode tubes. These simple timers initiate or interrupt the welding current at any point in the cycle and this is referred to as a non-synchronous arrangement. Where maximum versatility coupled with a high degree of accuracy is essential, it is usual to employ a fully synchronous system which includes power control by altering the firing point of the ignitrons. Modern electronic cubicles are designed to cater for spot, seam and stitch welding in one unit. The

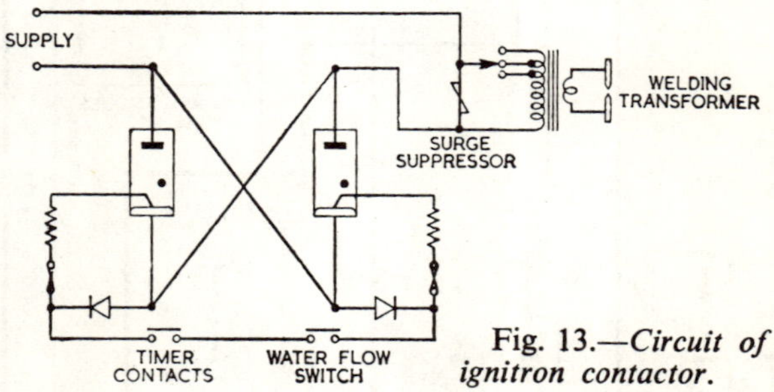

Fig. 13.—*Circuit of an ignitron contactor.*

introduction of metal into the welder throat is liable to cause unacceptable variations in welding current, and electronic methods are now available which enable the set value of current to be maintained. Certain welding applications necessitate a power control arrangement whereby the welding current increases at a controlled rate. This is termed "slope control" and necessitates special circuits coupled to the ignitron heat control system.

Rectification

The rectifier type of ignitron may be used in all the conventional controlled rectification arrangements. The conditions imposed on the tubes are considerably more arduous than when used as an electronic switch for resistance welding control. High peak inverse voltages are frequently encountered, and baffles and de-ionisation screens are included in order to remove any tendency towards arc back.

The circuit shown in Fig. 14 utilises three rectifier ignitrons in a conventional three-phase half-wave arrangement, and is capable of providing 600 A at about 300 V.

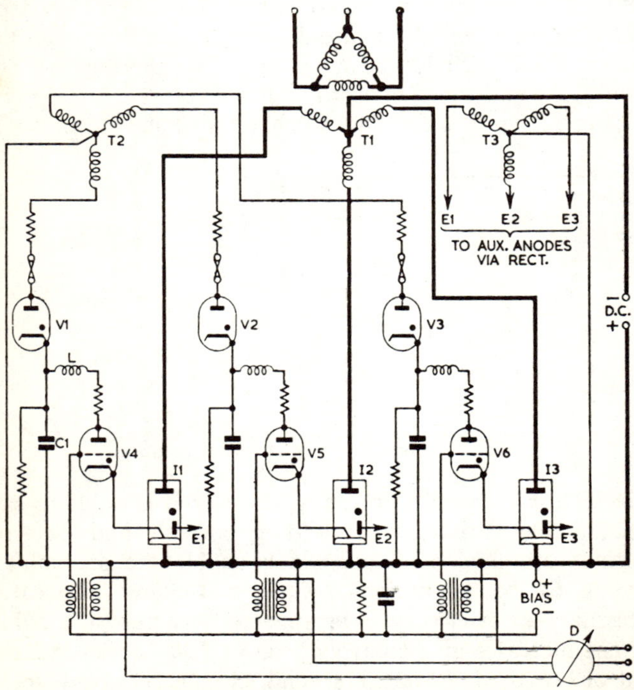

Fig. 14.—*Simplified circuit of an ignitron rectifier installation.*

The method of ignition is common to all three ignitrons and circuit references will be made to the components associated with I1. C1 will become charged through its associated rectifier (V1) when the anode of I1 is supplied with a negative half-cycle. The capacitor voltage, which is in the region of 400–600 V is now applied to the anode of thyratron V4 through an inductance L. The thyratron grid will be triggered at the appropriate point in the cycle corresponding to the degree of retardation required, and a current pulse of very short duration will flow through the ignitor, thus establishing the arc.

The auxiliary anode of the ignitron is supplied with power through a rectifier and its associated transformer T3. On no account must this auxiliary anode be supplied with continuously available direct current. This arrangement of the auxiliary anode ensures stable operation, and is particularly useful when a counter e.m.f. is produced by the load.

D is a phase-shifting device used for controlling the thyratron by means of an a.c. voltage superimposed on the d.c. bias.

Installation

These tubes are heavy current devices and all usual protective and isolation features should be included in the installation. Care must be taken to ensure that the ignitron is mounted strictly vertically.

The cooling water should be free of suspended matter and preferably not below 10°C. The water should enter the lower inlet, and the connecting hoses supplied must not be shortened, as the length chosen is such that the electrical resistance of the water column is adequate. Water flow switches are invariably included as a safety measure, the contacts being arranged to interrupt the ignitor circuit. When the conservation of the water supply is an important consideration, thermostats may be fixed to the water jacket and arranged to operate a solenoid water valve so that the water flow is related to the actual conditions prevailing.

Due to the abrupt cessation of the main current, transient voltages, which can be of quite a high value, may build up across the welding transformer primary. It is thus desirable to include either damping resistors or a surge suppressor. When used for welding service, the minimum current must not fall below 50 A otherwise the ignitor circuit may carry the load. Welding installations of low capacity should have a ballast resistor connected to bring the total load up to the prescribed value.

THE NEVITRON

This air cooled, single anode, mercury arc tube is a development of the multi-anode glass-bulb rectifier. It has several marked advantages relating in particular to the absence of a condensing chamber, a short arc path (resulting in greater efficiency), small dimensions and light weight.

One of the most significant features is the manner in which the cathode spot is anchored to the side of the pool, thus preventing random excursions over the surface which are characteristic of the normal pool rectifier. This type of tube can be operated in conditions where vibration, shock, and movement would prohibit the use of conventional systems and it is thus eminently suitable for mobile or shipboard use.

Construction

One of the major factors controlling the physical dimensions of a mercury arc rectifier is the question of mercury vapour-pressure within the bulb. This is related to the rate of evaporation of the mercury from the cathode pool and the heat dissipated in the arc. In the nevitron the cathode spot is anchored to a molybdenum cup containing the mercury pool, and this provides a convenient means of extracting heat from the arc by conduction. This adhesion occurs due to the fact that molybdenum is wetted by mercury, and a concave meniscus is created. Under full load conditions the cathode spot will form a complete circle around the periphery of the pool.

The excitation arrangements and general construction of this type of tube are shown in Fig. 15. At the instant of switch-

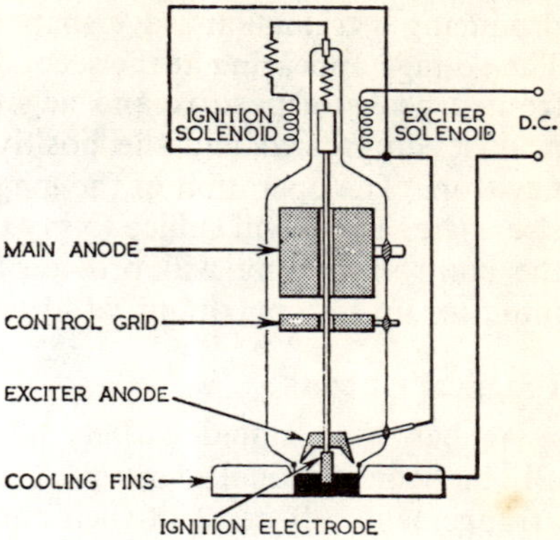

Fig. 15.—*Internal structure of a nevitron.*

ing on, d.c. is applied to the ignition solenoid through a path which includes the gravity-operated dipper, lifting it clear of the mercury pool, thus establishing the arc. The excitation current is arranged to flow through an additional winding so that the dipper is held clear of the pool as long as the arc is maintained.

Grid Control Arrangements

This type of tube is fitted with a control grid and all the usual controlled or uncontrolled rectifier arrangements can be utilised. Full grid control, either open or closed servo loop, is easily accomplished with small input signals, and this arrangement is widely used for variable-speed motor drive applications.

Reference has already been made to grid control schemes utilising a fixed amplitude a.c. signal superimposed on a varying d.c. level, and also a capacitor discharge arrangement operating in conjunction with a thyratron tube. The nevitron utilises a system consisting of a magnetic amplifier which supplies suitably phase-shifted firing pulses to the nevitron grids, and which includes a control winding whereby the servo loop may be closed.

These steep-fronted firing pulses are obtained from a peaking transformer delta-connected across the supply, each winding being in series with a reactor wound on a core of silicon iron producing a triangular wave shape from the sinusoidal input. The voltage appearing at the secondary is in the form of steep-fronted peaks of positive and negative polarity. The negative peak is suppressed and the positive peak is used to fire the nevitron. The operation of the magnetic amplifier is described elsewhere, and it will suffice to say that variations of current in the control winding will provide a phase shift of the firing impulses up to a maximum of about 130 electrical degrees.

Practical Circuits

As has been stated earlier, these tubes may be used in all the conventional controlled and free running rectifier arrangements. Because of their compactness and light weight, a very considerable saving in space so far as the entire installation is concerned may be accomplished if the main rectifier transformer can be avoided. By using the familiar three-phase bridge circuit arrangement, six single anode tubes may be used to provide full-wave, three-phase rectification, direct from the supply mains. If such an arrangement is to be used for supplying a d.c. motor from a 415 V three-phase supply the machine will have to be wound for a maximum of about 550 V. The output voltage (maximum conduction) can be calculated from the following expressions:

$$E_{\text{d.c.}} = 1{\cdot}35E_{\text{a.c.}} - 2E_{\text{arc}},$$

where $E_{\text{d.c.}}$ is the d.c. output voltage and E_{arc} the arc drop in the tube. Neglecting arc drop, the inverse voltage across any tube will not exceed $\sqrt{2}E_{\text{a.c.}}$.

A practical arrangement employing six 100 A nevitron tubes is shown in Fig. 16 and these are controlled by a magnetic amplifier.

Anode reactors must be used in each phase for protection purposes, chiefly as a means of limiting the short circuit currents which otherwise could reach large proportions.

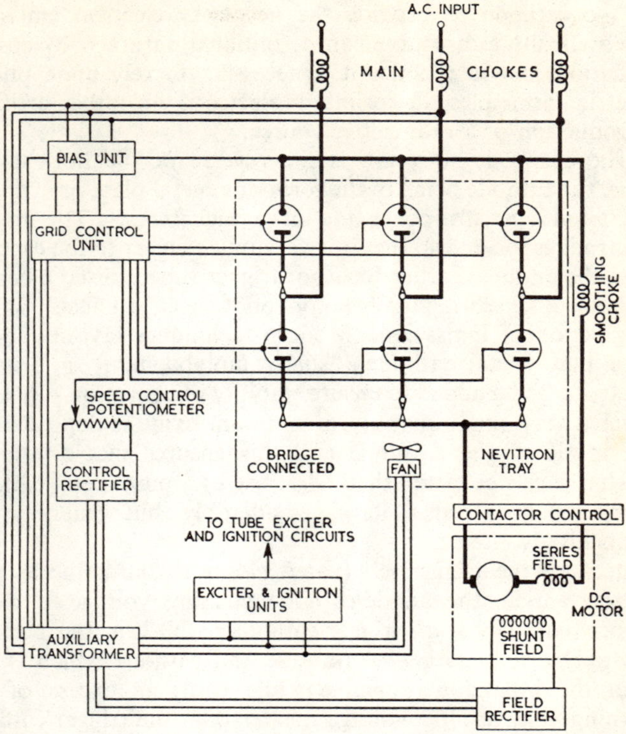

Fig. 16.—*Block diagram of bridge-connected nevitrons for direct-to-mains operation.*

Various auxiliary supplies are required and their position and function are indicated in the diagram. The motor field is of fixed value and supplied through metal rectifier units.

This type of variable-speed drive is capable of operating over a 60 : 1 speed range with a sensibly flat load/speed characteristic.

COLD-CATHODE TUBES

The cold-cathode tube does not employ a heated cathode or filament, relying instead upon the effect of ion bombardment of the cathode to provide the necessary electron emission. Theoretically a discharge can be initiated naturally by cosmic radiation, but in practice it is necessary to rely upon photo-electric principles, a priming electrode, or the artificial introduction of a radioactive source.

The electrode structure is the reverse of the hot-cathode tube, the cathode being in the form of a metal plate or cylinder, the anode usually consisting of a nickel rod. The trigger electrode is located so that it is very much closer to the cathode than the anode. Earlier forms of trigger tube utilised a coated cathode and relied largely upon photoelectric effects for the production of ionisation. Modern techniques favour the use of a pure metal cathode (usually molybdenum), a priming electrode being added to ensure stability of operation. In some tubes a very small quantity of uranium oxide is deposited on the inside of the envelope and this ensures that a priming discharge can be established. The use of a pure metal cathode raises the operating voltage considerably, but generally this is no disadvantage.

Because the trigger is situated closer to the cathode than to the anode, the anode/cathode striking voltage is much higher than that of the trigger/cathode path. It is thus possible to apply a voltage between anode and cathode which is less than the breakdown voltage while being in excess of the burning voltage. If a voltage greater than the trigger/cathode breakdown potential is applied to the trigger electrode, then a self-maintained discharge will occur over the larger anode/cathode path.

This type of tube is widely used in industry in conjunction with resistance-capacitance circuits to provide accurate time intervals varying from a fraction of a second up to ten minutes or longer. Typical examples include photographic exposure, resistance welding, diagnostic and therapeutic X-ray dosage, time of applied heat in induction and dielectric loss heating equipment. These are examples where only one or two timing periods are required. Cold-cathode timers may be used in cascade to provide a number of sequential or overlapping intervals. The more complex arrangements are used for determining the time cycle in die-casting machines, chemical processes, hydro-extractors, and in the refinement of sugar.

A simple two-stage timer is depicted in Fig. 17, the manner of operation being as follows. The normally closed contact RL1/1 ensures that C1 is in a fully discharged state at the commencement of the cycle. Upon initiation RL1/1 opens and RL1/2 closes, which enables C1 to charge exponentially at a rate determined by the series resistance. (It should be noted that the operating coil of the initiating relay RL1 is not shown.) When the potential across C1 has reached the level which causes a breakdown between trigger and cathode, V1 conducts energising RL2 which is then kept closed by the holding-on contact RL2/1. Due to the by-pass effect via R1 the tube will become extinguished. Additional contacts on RL2 will operate various electro-mechanical devices according to process

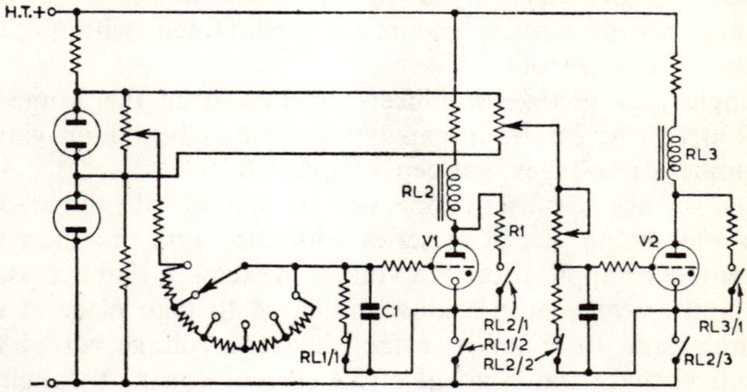

Fig. 17.—*Simplified circuit of a two-stage timing unit using cold-cathode tubes.*

requirements, at the same time closing the cathode circuit of V2 in order to initiate a precisely similar timing sequence in this stage.

Dekatrons

The dekatron, or cold-cathode counting tube, can count electrical impulses at very high speeds, displaying the total visually and, in addition, an output signal can be produced at a predetermined number. Industrial control applications include frequency division, electronic tachometry, high speed photoelectric counting and as an accurate digital timer in resistance welding processes. A number of different versions are manufactured to cater for various control requirements. These include types for double or single pulse counting, bi-directional computing, units suitable for scale of ten or twelve, and register tubes. The operation of these tubes is described in Section 8.

Stabilising and Reference Tubes

Both these tubes are cold-cathode gas-filled devices. Their functions, however, differ considerably. The voltage stabilising tube is required to provide a substantially constant voltage when the current through the system varies over wide limits. The primary duty of the reference tube is to provide voltage which can be accurately maintained within close limits for long periods.

Single gap voltage stabilisers operate over the range of 85–150 V, whereas multi-gap types are capable of providing a number of voltages between 70 and 280 V.

The voltage stabiliser is operated in parallel with the load, a resistance being placed in series with the supply to limit the current. The application of a voltage in excess of that necessary for ignition will cause a glow discharge to take place at the cathode, and once this is established the voltage across the tube is virtually independent of the current flowing through it. For satisfactory results as a voltage stabiliser the tube must be operated at current levels which are in the glow discharge region and the range available will be dependent upon the cathode area. For a given set of conditions an increase in current will result in the glow discharge covering a greater area of the cathode plate, and the current density will remain constant until the available space is fully utilised.

In error-actuated servo systems and other high accuracy measuring devices, a very stable reference source is imperative, and is frequently the limiting factor in the overall scheme. Comparisons with the zener diode in Section 3 should be made.

The voltage reference tube, although similar in construction to its stabilising counterpart, utilises special manufacturing techniques in order that maximum accuracy coupled with long term stability may be attained. The metal cathode must be free of impurities and the same stipulations apply in respect of the inert gas filling. The voltage reference tube must be operated at a constant and specified current level. If a voltage

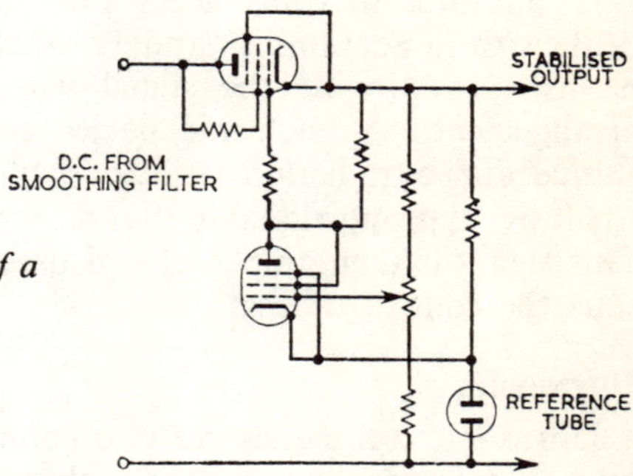

Fig. 18.—*Typical use of a voltage reference tube.*

reference tube is used in a circuit as shown in Fig. 18, a voltage accuracy of 0·1 per cent may be achieved.

SENSING SYSTEMS

The thickness or density of material, the level to which a container is filled and the proximity or alignment of objects are examples of things which can easily be controlled automatically by placing sensing elements in suitable positions and using the signals they provide to initiate control action. A clear example is the placing of a light source and photoelectric cell one above the other beside a reeling machine: if the material being reeled interrupts the light path between the light source and cell the resulting signal from the cell can be applied to equipment which will adjust the alignment of the material so that it is maintained nearly constant. A further example is one of many possible arrangements for automatically controlling the level of fluid, powder, etc. in a tank. If a capacitance probe is inserted in the wall of the tank at the point at which the level is to be maintained, when the material reaches this level the capacitance of the probe will change. This change can be detected and the signal thus derived amplified and used to operate an electro-magnetic relay which in turn may operate a system for automatically stopping the flow of material into the tank.

In addition to transducers and photo-electric equipment, dealt with in Sections 13 and 12, nucleonic devices may also be used to provide the signal which initiates the control arrangement. A nucleonic device consists of a radioactive source and a radiation detector which will produce a very small signal proportional to the amount of radiation it receives. This signal is then amplified and used as in other systems to start the control process.

Ultrasonics

Ultrasonic techniques are also commonly used. Ultrasonic waves (pressure waves above about 20 kHz) can be propagated in straight lines, focused and reflected, and ultrasonic systems make use of these properties. The waves are usually generated by means of a magnetostrictive or piezo-electric generator. In the former mechanical vibrations are produced by passing a.c. at the desired frequency through a winding surrounding a ferromagnetic core; in the latter an alternating voltage of the desired frequency is applied between two faces of a suitably cut crystal which is thereby forced into oscillation. The output of an ultrasonic transducer can be used with a suitable receiving transducer as the basis of a control system.

The energy produced by an ultrasonic generator can be concentrated by means of a lens or concave mirror (in liquids), an inverted horn (for gases) or a solid tapered member (for solids) to build up very high local energy intensities for particular applications. Practical applications of ultrasonics include blind guidance, depolymerization, dispersion of solids, elasticity measurement, emulsification, fatigue testing, flaw detection, smoke precipitation, ultrasonic cleaning, drilling and soldering, and ultrasonic therapy.

Servo Amplifiers

Where a servo amplifier is used to control directly the operation of a d.c. motor, the scheme shown in Fig. 6 (page 237) is generally used, with the input from a transducer or tacho-generator applied, via a phase-sensitive detector where changes of direction as well as speed are involved, to a phase-splitter or differential amplifier feeding a pair of power valves or transistors which control the current flowing in the two halves of a d.c. motor or generator with a split, i.e. centre-tapped, field winding. In view of the nature of the control signal direct connection between stages is generally employed, often with phase-advance capacitors in series with the coupling resistors at some point (these act in the same manner as the speed-up capacitors described in Section 8) to improve the response of the amplifier to a change in input. To improve stability, a low-pass filter may also be added (see phase advance + integral, Fig. 1 (page 232).

15

ELECTRO-MAGNETIC DEVICES

SOLENOIDS

A solenoid is a coil of wire wound round a cylindrical former: when a direct current is passed through the coil, the arrangement acts as a magnet. The electro-magnetic device known as

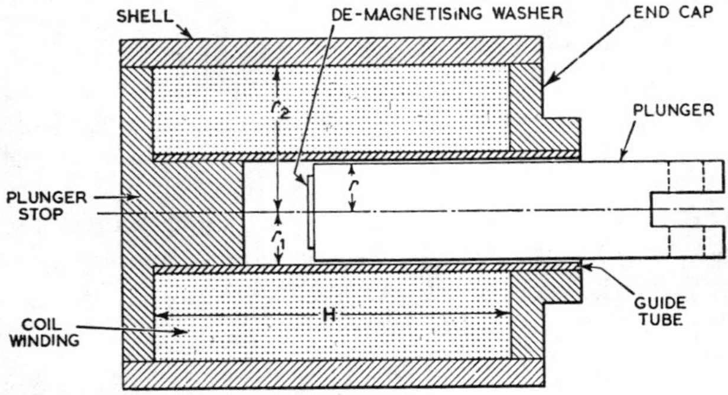

Fig. 1.—*Cross-section showing the basic components of a d.c. solenoid.*

a solenoid is used to convert electrical energy into mechanical energy. The simplest form consists of a cylindrical coil on a former within the bore of which an iron plunger is free to move. When a current is passed through the coil it acts as a magnet thereby controlling the movement of the plunger. The pull exerted on the plunger is greatly increased if an end stop for the plunger is provided to complete the magnetic path as far as possible. The basic components of a d.c. solenoid are shown in Fig. 1. The plunger, shell, stop and end cap are made of a magnetic material. The guide tube is made of non-magnetic material, e.g. brass or phosphor bronze. The coil former is a non-magnetic bobbin on which is wound a suitable number of turns of insulated wire. The shape of the end of the plunger has an important effect on the force obtained for a given stroke. To avoid the effect of residual magnetism when the coil winding is de-energised, a disc of non-magnetic material is fixed to the sealing face of the plunger or plunger stop to provide an air gap.

The maximum pull of a simple solenoid (i.e. with no end stop) is proportional to the magnetising force and to the cross-sectional area of the plunger, and is given in pounds by the following formula:

$$P = kAIT/L$$

where A is the cross-section of the plunger in square inches, L the length of the coil in inches, IT is the number of ampere-turns per inch and k is a constant the value of which depends on the proportions of the solenoid and the degree of magnetisation of the plunger (for normal designs it ranges between about 0·0057 and 0·0096). It is usual to use the small value of k where the actual value is not known thereby providing a conservative estimate of pull. In the case of a normal solenoid with an end stop, the attraction between plunger and end stop must be added. This pull is given by $A(IT/2660l)^2$, where l is the air gap in inches. The total pull is thus given by the formula

$$P = AIT(1/175L + IT/2660^2l^2).$$

When working with gravity the weight of the plunger must be added to the total pull.

A.C. Solenoids

In an a.c. solenoid the flux passes through a maximum and falls to zero twice in each cycle, so that the pull is not constant and the efficiency is much less than that of the d.c. solenoid. The same basic principles apply to the design of both a.c. and d.c. solenoids, but the construction is quite different. A typical a.c. solenoid assembly is shown in Fig. 2. A laminated plunger and frame are used to minimise the effects of eddy currents.

Each lamination is insulated on one side with a suitable material such as varnish or porcelain. The laminations are built up to the desired stack for the armature and yoke and are firmly riveted and often faced on both sides with thicker plates of mild steel. The mating surfaces need to be carefully

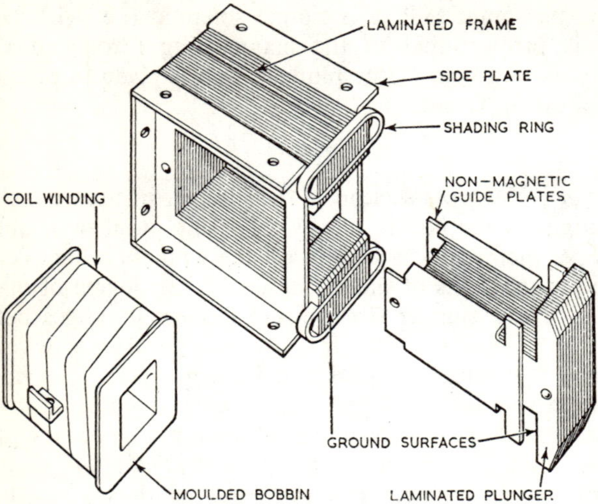

Fig. 2.—*Typical a.c. solenoid assembly.*

ground to ensure that the solenoid is not "noisy" and the air gap must be closely dimensioned as this can affect the VA input to the coil. To prevent noisy operation due to the a.c. component of force at twice the supply frequency, shading poles are fitted to one mating surface of the solenoid. This shading pole is normally fixed to the yoke on the mating surface, although sometimes it is incorporated in the top of the plunger. The guide tube is of non-magnetic material and is often slotted to reduce eddy current.

As the d.c. solenoid is more efficient than the a.c. solenoid it is often preferable to use a d.c. solenoid with a.c. supplies by incorporating a rectifier.

RELAYS

An electro-magnetic relay is fundamentally an electric switch designed to open and/or close a circuit when the current through its operating coil is made to flow or is interrupted or, in a few cases, is increased or decreased in magnitude.

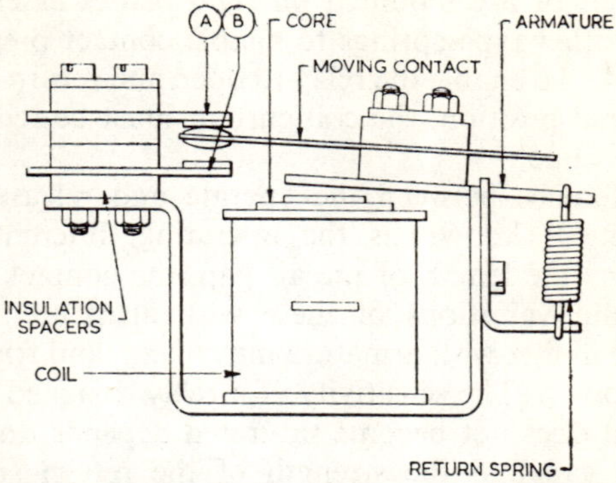

Fig. 3.—*Conventional change-over or double-pole relay.*

A small current through the relay coil enables a much greater current in a separate circuit to be controlled. Thus a relay may serve to pass on a weak signal with renewed or increased energy. Also as the operation of a relay may be delayed (by "slugging") it may be used for timing purposes to introduce a delay into a system.

A conventional change-over or double-throw type relay is shown in Fig. 3. When the coil is not energised the magnetic force does not overcome the spring-restoring force and the moving contact and fixed contact A are in contact, forming a "normally closed" circuit. Upon increase of the coil current to the "operate" value, the flux in the magnetic circuit increases to a level whereby the magnetic force begins to overcome the spring-restoring force, and the armature tends to move, reducing the air gap. With the decrease in air gap, the reluctance of the magnetic circuit also decreases, resulting in an increase of flux and hence force. Although the spring-restoring force increases, it is not at a rate sufficient to prevent the armature closing. When the armature closes the moving contact and fixed contact B engage. They are known as "normally open" contacts. The contacts are either attached to spring blades of beryllium copper, phosphor bronze or nickel silver, or are mounted on rigid blades assembled with helical or other type springs to enable contact pressure to be maintained. To enable the relay to open and return the contact to its normal position, the coil current must be reduced to its "release" value.

The difference between the operate and release values of coil current is known as the operating differential and is governed by the length of the air gap, the contact and spring pressure and variations of these with armature movement, the change in flux with armature movement, and friction in the armature pivot. The sensitivity of a relay designed so that the iron circuit does not become saturated depends on the length of the air gap and the strength of the return spring. The operating speed of a relay is a function of the size of the moving parts, the spacing of the contacts and the electromagnetic inertia. High-speed operation calls for light mechanical parts and low-inductance circuits.

For use with alternating currents, it is essential to incorporate a shading pole, which provides eddy-current damping of the armature.

Types of Contact

The four basic types of contact are the normally open, which close with operation of the relay (M type); the normally closed, which open with relay operation (B type); the changeover, in which the moving contact is common and breaks from the normally closed contact before making with the normally open contact (C type); and the make-before-break which are changeover types that make with the normally open contacts before breaking with the normally closed ones (K type). These are illustrated in Fig. 4.

Apart from the simple on/off relay, there are also multi-contact types and the "hold on" type in which the contacts are so arranged that after an initial signal the contacts close

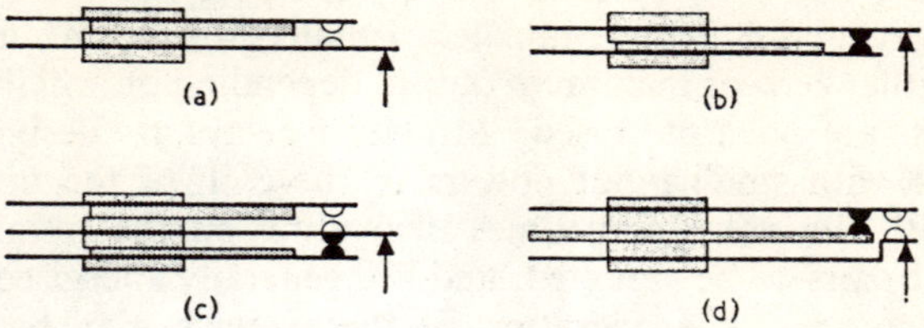

Fig. 4.—*Basic types of relay contacts.* (a) *Normally open* (*M type*). (b) *Normally closed* (*B type*). (c) *Break-before-make* (*C type*). (d) *Make-before-break* (*K type*).

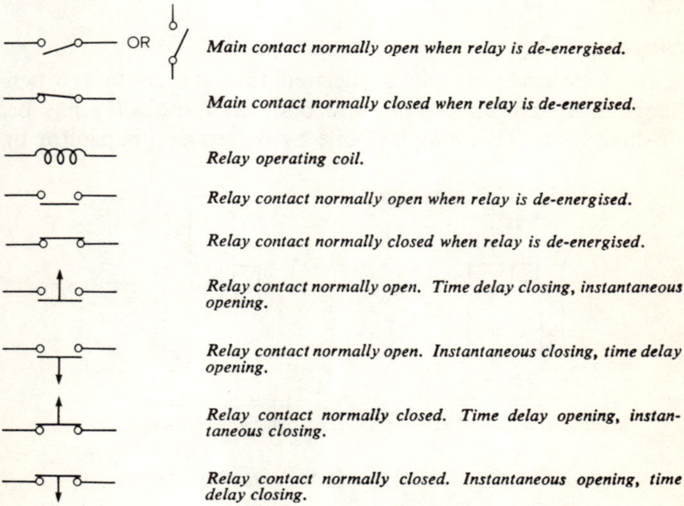

Fig. 5.—*Basic symbols used in relay circuit diagrams.*

and remain closed until released by a further signal. The basic symbols used in circuits incorporating relays are shown in Fig. 5.

Miniature Relays

The miniature relay may be regarded as a two-state amplifier having one input signal, supplied to the magnet coil, and one or a number of output signals. These output signals, which can be completely isolated one from another, have either zero or maximum output depending on whether the contacts are open or closed. Miniature relays are designed to operate with small input powers to the coil, of the order of 0·1–2 W, the value sometimes depending on the number of contact pairs to be actuated, and are generally energised from a d.c. source. The operation of these relays may be made extremely sensitive by limiting the armature movement. This will, of course, reduce the contact gap and hence the switching capacity.

Slugging Relays

In some cases it may be desired that the contacts open a short time, e.g. up to half a second, after the relay has been de-energised. This may be done by means of a capacitor or a

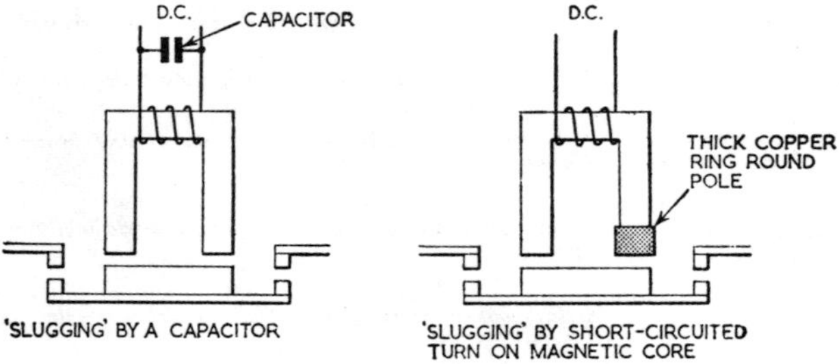

Fig. 6.—*Two methods of "slugging" relays.*

short-circuited turn on one of the magnetic poles. These methods are illustrated in Fig. 6. The short-circuited turn provides a secondary magnetic field which holds the contact closed for a brief period after the main relay coil has been de-energised.

Polarised Relays

A polarised relay is one incorporating a magnet providing a permanent flux. When the operating signal arrives, the flux this produces counteracts the permanent flux in the relay and the contact is attracted away from its normal position. The contact gap can be made very small in polarised relays, and they are therefore capable of extremely rapid response. If the permanent magnet is suitably arranged, the normal position of the moving contact can be "neutral" between the two fixed contacts.

Transistorised Relays

It is often required that a relay switching power loads should operate from an input of a few milliwatts. This often cannot be achieved with polarised or sensitive relays. A

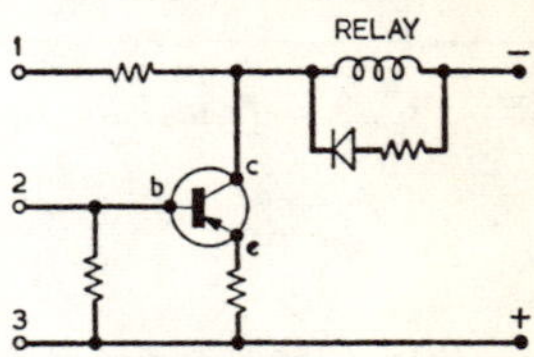

Fig. 7.—*Transistorised relay circuit.*

general-purpose relay may, however, cope with this requirement if the coil-operating power is suitably amplified. This may be done by including a transistor amplifier in the circuit as shown in Fig. 7. Large voltage transients can occur when the relay switches, and to prevent damage to the transistor the diode with series resistor are connected in parallel with the relay coil. The emitter resistor provides stability due to the negative feedback it introduces. Its value would be 100 ohms or so, the value of the base resistor being about 1,000–2,000 ohms. When terminals 1 and 2 are connected, the relay operates; connecting terminals 2 and 3 by some external means makes the relay drop out; the amplifier is sometimes built into the same case as the relay itself.

The response speed of the relay is not greatly increased by the use of transistor amplification. Where greater coil operating power is required a second transistor can be connected in cascade to provide further gain.

Various other circuit arrangements provide such features as triggered relays responding to pulses and relays with small time delays in operation. An example of the latter is shown in Fig. 8: the capacitor is charged through the fixed (22k) and

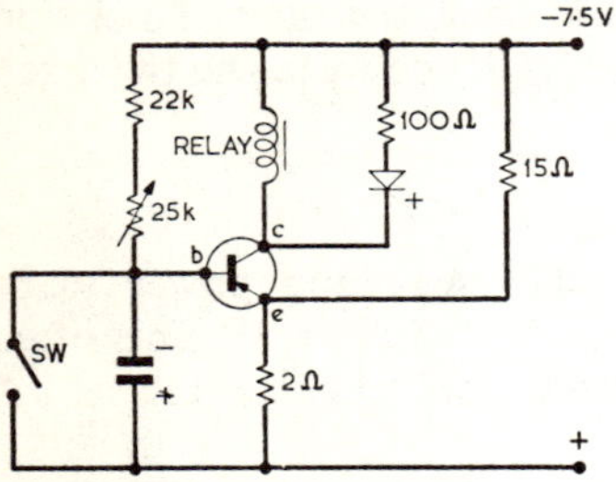

Fig. 8.—*Transistor timing circuit.*

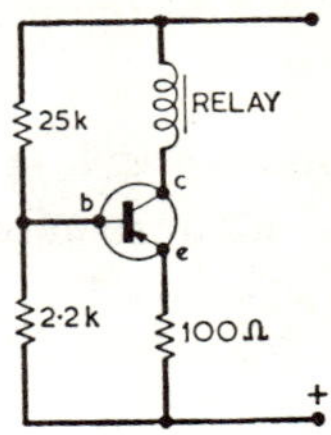

Fig. 9.—*Method of stabilising a transistor against "thermal runaway".*

variable (25k) series-connected resistors when the switch is opened. Delays of up to one second can be obtained with this circuit.

Relay units generally employ germanium transistors for which the maximum operating temperature is about 70°C. Allowance must, however, be made for a rise in the junction temperature caused by internal heating in the transistor. This temperature rise is about 0·4°C/mW, e.g. for a transistor operating at 100 mW the rise in temperature would be 40°C. Thus maximum ambient temperature must equal the maximum operating temperature minus the temperature rise, i.e. in the example mentioned this would be 70° − 40° = 30°C.

The base-to-emitter voltage in a transistor is dependent on temperature and falls by about 2·5 mV/°C. This causes an increase in base current of the order of 10 μA and, for a transistor with a gain of 100, causes the collector current to increase by 1 mA. This increased collector current will cause a corresponding increase in the junction temperature. The effect is cumulative and can cause a "thermal runaway" resulting in damage to the transistor. A preferred way of stabilising against a dangerous rise in collector current due to increase in temperature is to use a potential divider and emitter stabilisation as shown in Fig. 9. An increase in emitter current will produce an increased voltage drop across the emitter resistor decreasing the voltage applied between the base and emitter. This produces a drop in the base current, resulting in a fall in collector current. This reduces the self-heating of the transistor, thus reducing the original temperature rise.

Microminiature and Dry-reed Relays

Microminiature relays small enough to be enclosed in an 0·37 in. × 0·4 in. standard TO-5 transistor can are now available with single changeover, two-pole changeover contacts or as latching relays. With a voltage range of 1·3 to 30 V d.c. and operate or release times of 2 ms, they are suitable for switching at speeds up to 20 Hz. Switching capacity is up to 0·5 A at 28 V d.c. Life expectancy is of the order of one million operations.

A type of miniature relay that has come into wide use (including P.O. electronic telephone exchanges) is the dry-reed relay. This basically consists of a sealed insert containing two flat magnetic alloy cantilever reeds hermetically sealed into opposite ends of a narrow glass tube (typical insert lengths 1 to 3 in.); the ends of the reeds overlap, typically by about 0·015 in., and are separated by a gap of about 0·005 in. The overlap may be gold plated or diffused to provide suitable switching characteristics and reduce any tendency to stick.

The insert is mounted within the operating coil, which when energised results in the reed contacts closing. Several inserts may be operated by a single coil. Alternatively, switching can be controlled by mechanical movement of a permanent magnet. Larger sizes may switch up to about 250 mA, and smaller ones about 100 mA. Typical operate time is about 2 ms, and release time less than 100 μs. This type of relay is essentially a low-power device mainly confined to make or break switching but at much faster switching times than conventional relays.

Ferreed Relays

A particular form of reed relay which has been developed for use in large electronic telephone exchanges in the United States comprises a combination of ferrite and magnetic reed switch; in practice a magnetic alloy called Remendur is usually substituted for the ferrite.

The Remendur can be switched in a matter of microseconds between the two magnetic states by means of very short current pulses; it has a square hysteresis loop and can be switched with relatively low power.

MAGNETIC AMPLIFIERS

The magnetic amplifier is a useful control device which is both robust and reliable. The basic element in a magnetic amplifier is the "saturable reactor", also known as the "transductor". This, in its simplest form, consists of a core of electro-magnetic material on which are wound two windings, one carrying alternating and the other direct current. The a.c. winding is connected in series with an a.c. supply and a load, the current through which is to be controlled. By varying the current in the d.c. winding the a.c. impedance of the circuit can be changed and in this way the current through the load is controlled. With suitable windings a small change in direct current can be made to produce a large change in the load current (since the d.c. change can vary the degree of magnetic saturation of the core), and in this lies the usefulness of the device. As the change in load current is normally greater than the change in control current, a saturable reactor by itself forms a simple current amplifier. The term magnetic amplifier, however, is usually considered to imply a circuit arrangement incorporating one or more saturable reactors and other circuit elements such as rectifiers, and sometimes having additional windings on the reactors to improve, for example by means of feedback, the characteristics of the amplifier.

Basic Circuit

A simple magnetic amplifier circuit is shown in Fig. 10. The windings of two identical transformers are arranged back to back so that no voltage, at the supply frequency, appears between the terminals A and B. The only current in the a.c. circuit is the magnetising current which the transformers will draw even if a low impedance is connected across AB. If a direct current is passed from B to A through the windings the cores will run into partial or complete saturation. The impedance they offer to the a.c. supply voltage will thus fall and the current in the a.c. circuit will rise. In this way d.c. in the additional set of windings controls the a.c. in the load.

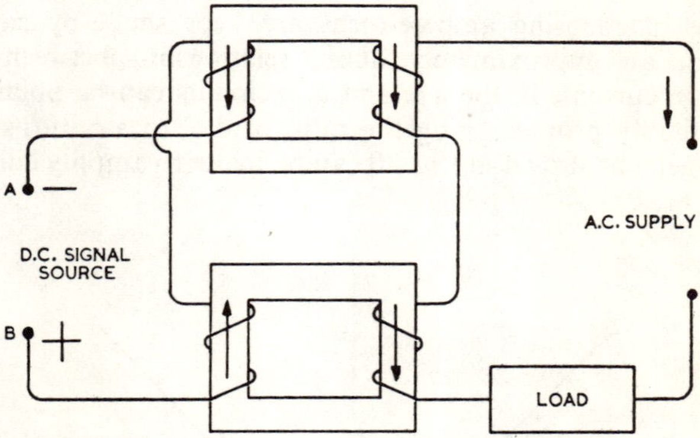

Fig. 10.—*Basic magnetic amplifier circuit.*

The a.c. in the load is rich in odd harmonics and even harmonics appear in the d.c. circuit. If the mean values, taken over a half-cycle of the supply voltage, are measured it is found that:

a.c. ampere-turns

$$= \text{d.c. ampere-turns} + \text{magnetising ampere-turns.}$$

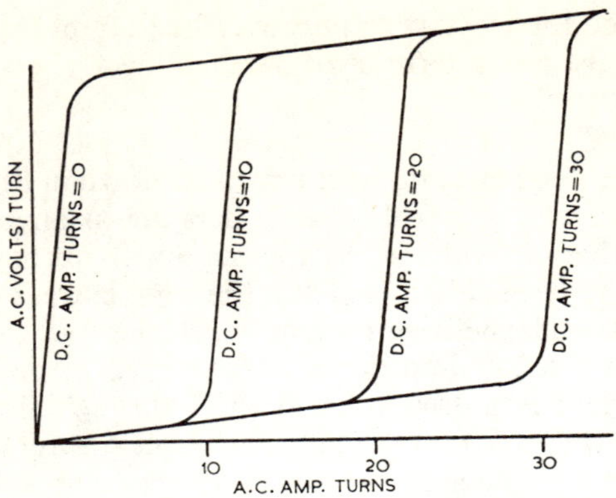

Fig. 11.—*Characteristic curves of a.c. output against applied voltage for different values of d.c. in the control windings.*

If the magnetising ampere-turns are kept small by careful design, an approximately linear relationship between the average currents in the a.c. and d.c. circuits can be obtained. By suitably proportioning the ratio of d.c. to a.c. turns, the arrangement shown in Fig. 10 can be made to amplify current

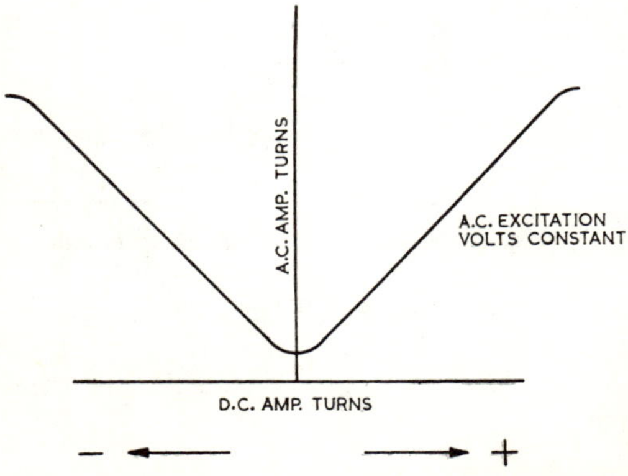

Fig. 12.—*Characteristic curve of a.c. output against d.c. input.*

and hence power in a manner substantially independent of variation in the voltage and frequency of the a.c. supply or in the resistance of the a.c. circuit. Figs. 11 and 12 illustrate these characteristics. Current gains of 10 and power gains of 40 are readily obtainable with this circuit.

Circuit Using Positive Feedback

Higher gains can be achieved by the incorporation of positive feedback. Fig. 13 shows the arrangement. Some loss of stability owing to supply voltage and frequency variations is inevitable with this circuit, but current gains of 200–300 and power gains of 10^4–10^5 are quite easy to obtain. The waveforms of the currents in the a.c. and d.c. circuits are substantially the same as with the simple arrangement previously considered. The

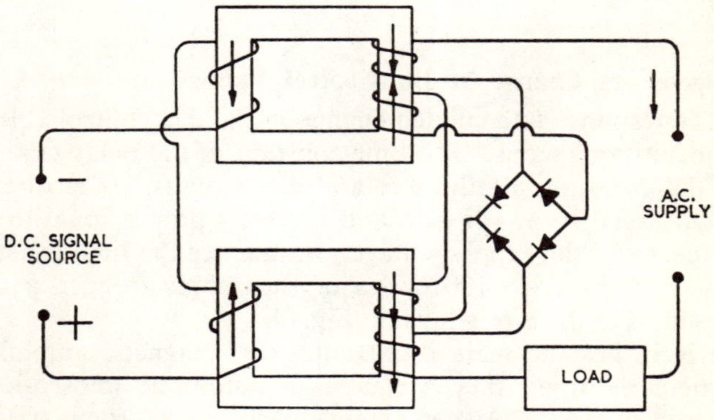

Fig. 13.—*Magnetic amplifier circuit incorporating positive feedback. The positive feedback increases the amplification and makes the amplifier sensitive to the polarity of the d.c. source.*

positive-feedback magnetic amplifier is, however, sensitive to the direction of the controlling direct current, as can be seen from the characteristic shown in Fig. 14. In practice biasing ampere-turns are often used to move the operating point to the middle of the linear portion of the characteristic, in a similar manner to that employed in thermionic valve amplifiers.

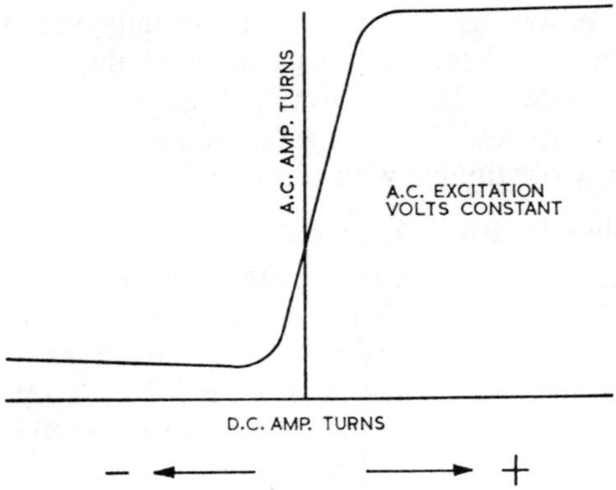

Fig. 14.—*Characteristic of a magnetic amplifier incorporating positive feedback.*

Response to Change in d.c. Control Voltage

The response to a sudden change in the d.c. control voltage is not instantaneous. The time-constant of the delay depends on the constants of the d.c. and a.c. circuits. It is directly proportional to power gain and inversely proportional to the frequency of the supply voltage. In practice the time-constant may be from 2 to 100 cycles of the supply frequency with amplifiers of the type shown in Fig. 13.

Herein lies the main disadvantage of magnetic amplifiers, particularly when they are used in automatic (closed-loop) control systems. Delays of this order make such systems sluggish in response and difficult to stabilise.

Further Variations

To achieve greater gain, the basic arrangements shown in Figs. 10 and 13 can be connected in cascade. These basic arrangements can also be connected in push–pull, giving increased stability and providing a reversible d.c. output.

A comparison of the basic thermionic valve and magnetic amplifier circuits is shown in Fig. 15.

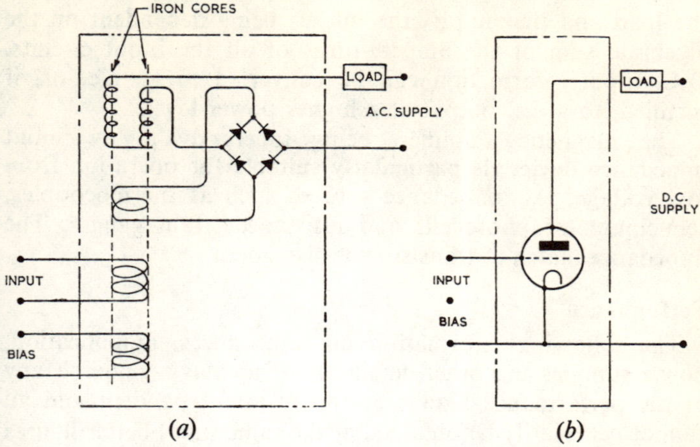

Fig. 15.—*Comparison between* (*a*) *basic magnetic and* (*b*) *thermionic triode valve amplifier circuits.*

General Characteristics

Unlike the thermionic valve amplifier, which is a voltage-operated device having greatest efficiency when working from a high impedance source, the magnetic amplifier is a current-operated device requiring a comparatively low impedance source rather like the transistor amplifier.

The magnetic amplifier contains no moving parts, is quiet, can be hermetically sealed and has an almost unlimited life. Its size is about three times that of a transformer having the same power output, varying, as with transformers, inversely with the supply frequency. As there is no warming-up time, it functions as soon as it is switched on. Unlike active devices, whose impedance is ohmic producing considerable internal power loss, there is, because their impedance is mainly reactive, very little internal dissipation in magnetic amplifiers.

The input and output circuits of a magnetic amplifier are completely isolated electrically. Several input circuits can be provided, each electrically isolated from the others and from the load and the supply, the output being dependent on the algebraic sum of the ampere-turns of all the input circuits. D.C. input information can be converted to an a.c. or, if rectified, to a d.c. output at a higher power level.

The magnetic amplifier, being inherently a low input impedance device, is particularly suitable for operation from low voltage, low impedance sources such as thermocouples, semiconductor photocells and resistance strain gauges. The impedance match to transistors is also good.

Performance

The following information on impedances, amplification, power supplies and other details is intended as a guide. Many of the performance figures given are interdependent and all cannot necessarily be obtained at the same time; better figures can often be obtained under special circumstances.

The power supply is normally single-phase a.c.; a transformer is usually required to provide the correct voltage to the amplifier. Frequencies of 50–2,000 Hz are normally used; the effect of supply frequency on size and response time has already been noted.

The input may be d.c. or low frequency a.c. The input resistance is normally in the range of 0·1–3,000 ohms. The output can be d.c. or a.c. into impedances of 10–2,000 ohms using a 50 Hz supply. Higher frequency supplies allow proportionately higher values to be used. The d.c. output can be made reversible by push–pull techniques and an a.c. output can be made to change phase by 180 degrees.

Single-phase power gains of up to 10^4 are common, and amplifiers are possible with gains of 10^6 in one stage; higher gains can be obtained by cascade operation of two or more stages. With applications to low power amplification, the drift under normal operating conditions (including ± 10 per cent supply-voltage and ± 5 per cent supply-frequency variations) is equivalent to an input power of about 10^{-12} W. Linearity and stability are usually between 0·5 and 10 per cent of the maximum output current.

Response Time

It has been pointed out previously that the rather long time-constants of magnetic amplifiers when responding to sudden changes in the d.c. control voltage can be a disadvantage in some applications. Various methods of reducing the time-constant have, however, been devised. For example, the time-constant is inversely proportional to the frequency of the supply. Thus, driving the amplifier from, say, a 500 Hz instead of a 50 Hz supply will reduce the time-constant to one-tenth of its value at 50 Hz. Owing to the difficulty of obtaining such frequencies by other than electronic means, it is seldom practicable to work at frequencies higher than 1,000–2000 Hz. Again, as the time-constant varies directly with gain, reduction of gain reduces the time-constant; since, however, gain is usually what is required, this method is somewhat limited in application. The gain should, however, always be kept as low as practicable.

Cascade Circuits

When two or more magnetic amplifiers are arranged in cascade the overall time-constant is approximately equal to the sum of the separate time-constants while the overall amplification is the product of the separate amplifications. Thus a shorter time-constant for a given amplification can be obtained by employing two or more stages in cascade.

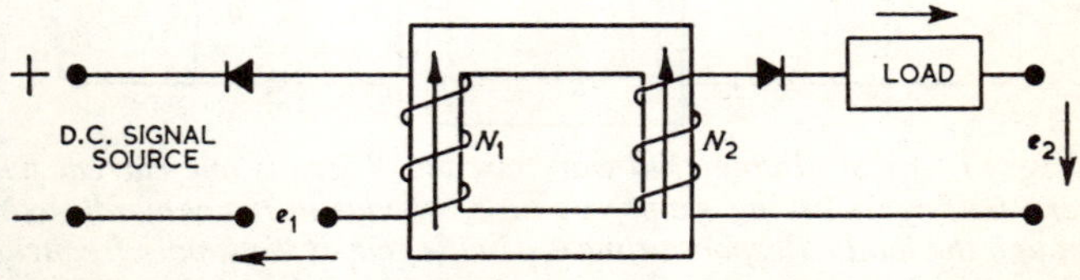

Fig. 16.—*Basic circuit of the Ramey half-wave magnetic amplifier e_1 and e_2 are a.c. supply voltages of the same frequency and of the phase shown by the arrows. Their amplitudes are such that $e_1/e_2 = N_1/N_2$ where N_1 and N_2 are the turns of the signal and load windings respectively. There is one current pulse in the load every cycle of the supply voltage. Response time is one cycle of the supply frequency.*

In 1951 Ramey introduced a magnetic amplifier circuit which responds within one cycle of the supply frequency but with sacrifice in power gain. Fig. 16 shows the basic circuit of this type of amplifier, which gives power gains of the order of 100–200 as against 10,000 for the type previously described. The basic arrangement shown in Fig. 16 only gives a half-wave output; Fig. 17 shows the Ramey circuit developed to give full-wave output.

Where power gain can be sacrificed for fast response, or where a multi-stage amplifier is acceptable, the high-speed circuits shown in Figs. 16 and 17 should be used. Where, however, a long time-constant can be tolerated, for example in control systems having comparatively long time-constants themselves (process control), it is usually advantageous to use the circuit shown in Fig. 13 in view of the high power gain that can be obtained with this.

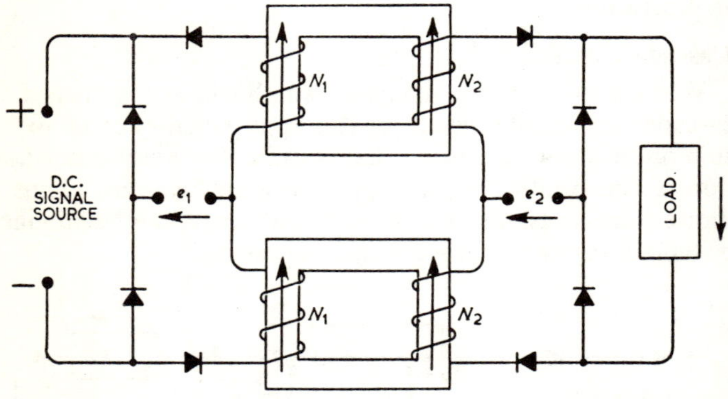

Fig. 17.—*Basic Ramey full-wave circuit. There is one current pulse every half-cycle of the supply voltage, always in the same direction, through the load. Response time is a half-cycle of the supply frequency.*

Applications

Magnetic amplifiers may be used to amplify the signals from such sources as semiconductor photocells, thermocouples and high frequency thermo-junctions, resistance strain gauges, inductive pick-ups, synchros, etc. The amplified output may be fed into heaters, lamps, relays, indicating and recording instruments, and a.c. and d.c. motors with field or armature control.

Some typical applications are: servo systems where magnetic amplifiers are used with d.c. split-field or a.c. split-phase motors and control fields of Metadyne and Amplidyne generators, temperature recording, control and alarm, recording of gas flow, photoelectric counters, voltage and frequency control of machines, automatic battery charging, automatic control of street lighting and sensitive relays.

16

POWER SUPPLIES

A.C. RECTIFICATION

Whereas a wire or other normal conductor presents the same resistance to current flowing in either direction, there are other circuit elements which offer a much greater resistance to current flow in one than in the other direction. If a non-linear element of this type is introduced into an a.c. circuit it will result in the average current during one half-cycle being greater than during the other half-cycle. A non-linear device which may be unidirectional or asymmetrical can thus be used to rectify a.c., since a d.c. component appears in the circuit.

In practice the non-linear element is usually a thermionic or semi-conductor diode (see Section 2). Such a device in an a.c. circuit will produce d.c. of a pulsating nature, and this can be smoothed into pure d.c. by means of a ripple filter. The design of the ripple filter will determine to some extent the regulation of the supply, that is to say the difference in output voltage from zero to full load. Where necessary the regulation can be further improved by electronic regulation, using such devices as thyratrons, voltage regulator and stabiliser tubes, zener diodes as well as conventional thermionic valves.

Thermionic Diode

The resistance offered by a thermionic diode to a current flow from cathode to anode is relatively low, while the resistance in the reverse direction is extremely high. When connected in an a.c. circuit, current will flow during the half-cycle when the anode is positive with respect to cathode, but there will be negligible current flow during the half-cycle when the cathode is positive with respect to anode.

In its simplest form the thermionic diode has a heated filament, which forms the cathode, surrounded by an anode electrode to receive the electrons. In practice, most rectifier valves have indirectly heated cathodes consisting of a circular or rectangular sleeve having a coating of the oxides of thorium and barium as the emitting surface. The cathode sleeve is

Fig. 1.—*Anode current/voltage characteristic curve of a typical thermionic diode rectifier.*

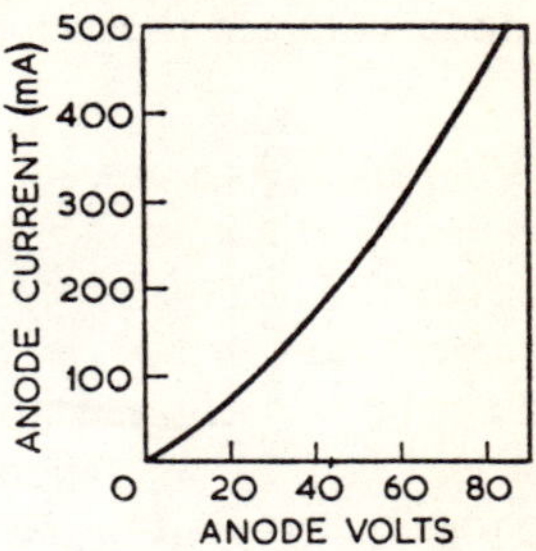

maintained at about 800–1,000°C by an internal heater and insulated from it by a spacer of alumina.

The indirectly heated cathode has a uniformly distributed emission and thus a longer life than the directly heated filament, with which the potential relative to the anode varies along its length.

The anode electrode is basically a tubular or rectangular tube, generally of nickel or nickel-plated iron, mounted concentrically around the cathode, and usually provided with ribs or fins to stiffen the structure and to increase heat dissipation. Anode temperature must be kept below that at which electron emission would occur, as otherwise electrons would flow in both directions.

Gas-filled Rectifiers

The high vacuum rectifier has appreciable internal resistance to current flow in the forward direction (that is electron flow from cathode to anode), due mainly to the mutual repulsion of the negatively charged electrons in the space surrounding the cathode. This internal resistance can be reduced by having a supply of positive ions in the cathode anode space to neutralise the electrons. This effect can be achieved by introducing a trace of inert gas into the valve, the required positive ions being produced by the collision of fast-moving electrons with the gas molecules.

The inclusion of gas modifies the anode current/anode voltage characteristics of the rectifier (see Fig. 2). At very low anode voltages the velocity of the electrons is insufficient to

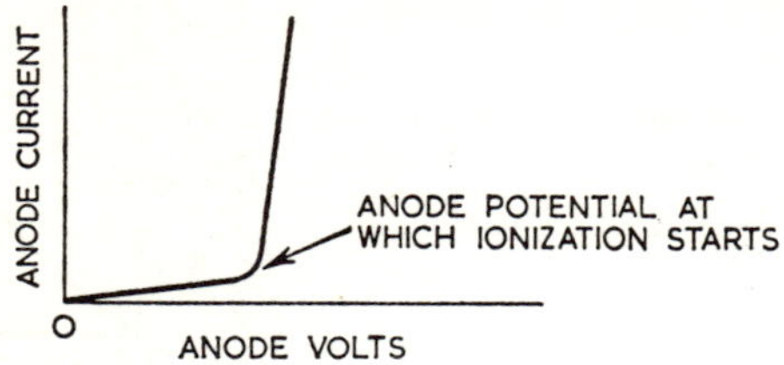

Fig. 2.—*Typical anode current/voltage curve of a gas-filled rectifier valve.*

produce positive ions by collision. However, if the anode voltage is increased until the point at which ions are produced, the space charge is neutralised and the internal resistance of the valve drops rapidly. The potential at which ionisation occurs varies with the type of gas: e.g. mercury 10·4 V, xenon 11·5 V. When ionised a characteristic glow discharge is visible.

Metal Rectifiers

For many years before the intensive development of germanium and silicon semiconductors took place, use was made of copper-oxide and selenium barrier-layer rectifiers, generally termed "metal rectifiers". These consist essentially of a good conductor separated from a semiconductor by a very thin insulating layer. The rectifying properties depend upon the fact that a greater number of electrons can pass across the barrier layer from the conductor than from the semiconductor in the same unit of time. Such a device thus offers different opposition (i.e. resistance) to the passage of electric current in opposite directions.

In the copper-oxide rectifier, the elements consist of a plate of pure copper on which is formed a layer of cuprous oxide

(Cu_2O). Copper oxide rectifiers are used mainly where high and stable reverse resistance is necessary, e.g. in test equipment. In the more widely used selenium rectifier the semiconductor consists of a layer of selenium deposited on a metal support of iron or aluminium: the good conductor is formed by spraying the surface of the selenium with a low-melting-point alloy, leaving narrow rings unsprayed at the centre and circumference to avoid electrical contact between the good conductor and the metal base support.

Originally the reverse voltage across a single selenium rectifier unit was limited to about 20 volts, but units were later introduced rated at up to about four times this figure. This greatly reduces the weight and size of a rectifier since it now comprises many less units.

The working life of metal rectifiers is determined almost entirely by the operating temperature of the elements, this being governed by the electrical loading, the cooling and the ambient temperature. To assist cooling, the elements are spaced or fitted with cooling fins or clamped to the equipment chassis which acts as a "heat sink". Metal rectifiers operate at much lower surface temperatures than valve rectifiers, and therefore must dissipate heat by convection and conduction rather than radiation. It is thus necessary to ensure an unrestricted flow of cool air. The forward resistance of a metal rectifier tends to increase with age.

Junction Rectifiers

An ideal rectifier would offer zero impedance to current in one direction and infinite impedance in the other. A germanium or silicon pn junction power rectifier approximates to this definition, having extremely low forward resistance and extremely high reverse resistance. A single germanium cell may drop only 0·5 volt at a forward current of 100 amps, with a reverse current of only a few milliamperes: in consequence the power loss is small and the rectifying efficiency very high.

As with selenium and copper-oxide rectifiers, rectification takes place at a junction between dissimilar materials, but the junction of a germanium or silicon rectifier exists inside a single crystal of the semiconductor (see Section 3). This results in much higher permissible current densities. The forward resistance does not increase with age.

Germanium junction rectifiers are generally used for high currents and may have cooling fins fitted. Silicon rectifiers have a higher reverse rating characteristic and operate at higher temperatures. Junction rectifier failures are generally caused by voltage surges beyond the maximum permissible peak inverse voltage. These surges are most likely to occur when the circuit is switched on or off, or from overvoltages appearing on the mains supply.

Series operation of several rectifiers of the same type may usually be arranged by reducing the peak inverse voltage rating and shunting each with an equalising resistance.

Thyratrons and Silicon-controlled Rectifiers

For industrial control where voltage control is required—for example to operate a variable-speed d.c. motor by supplying variable d.c. voltages to the motor armature and field circuits—the thyratron, or its semiconductor equivalent the silicon-controlled rectifier, may be used. The operation of these devices is described in Sections 3 and 14.

RECTIFIER CIRCUITS

There are a number of ways in which rectifiers may be connected into circuit, the choice being largely governed by the d.c. voltage required, the nature of the load (which may be purely resistive, resistive/inductive, resistive/capacitive or almost any combination of resistance, inductance and capacitance), the degree of ripple reduction required, and whether the supply is single- or three-phase. Either thermionic or semiconductor diodes may be used in these circuits, but in the case of thermionic diodes provision must also be made for the supply of suitable voltages for the heaters. In the following description the rectifier is assumed to be isolated from the mains supply by means of a double-wound transformer, although this may not always be the case.

Single-phase Half-wave Rectifier

The simplest circuit is that shown in Fig. 3. Current from the transformer secondary flows through the rectifier and into the load only on alternate half-cycles. With a truly resistive

Fig. 3.—*Single-phase half-wave rectifier with resistive load.*

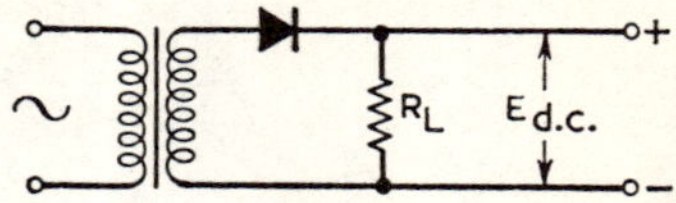

load, the load current and the load voltage will comprise half sinusoids of similar waveform to the positive half-cycles of the transformer secondary voltage, less the voltage drop in the rectifying element. During the negative half-cycle the peak inverse voltage across the rectifying element is equal to that of peak a.c. supply voltage.

Reservoir Capacitor

The addition of a capacitor across the load (Fig. 4) profoundly modifies the mode of action. Assuming that steady conditions are established in the circuit, there will be a d.c.

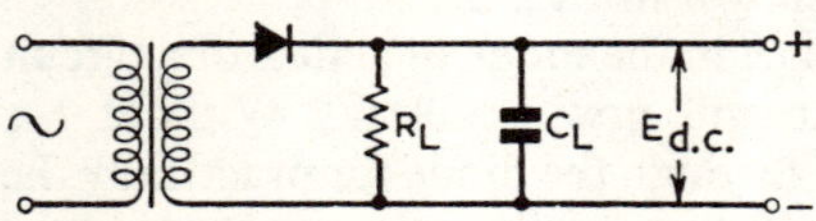

Fig. 4.—*Addition of a reservoir capacitor to a half-wave rectifier circuit.*

voltage across the load resistance R_L and the parallel capacitor C_L. Current will not flow through the rectifier diode into the load until the instantaneous value of the transformer voltage on the positive half-cycle exceeds the d.c. voltage on the capacitor, that is: $E \sin \theta > E_{d.c.}$. Current then flows through the diode into the load resistance and capacitor in parallel, and the voltage across these rise with the rising 50 Hz voltage up to the peak of the a.c. cycle. It then commences to fall with the supply voltage, but at this point the capacitor commences to discharge into the load resistance, and the rate at which the capacitor voltage falls is controlled only by the exponential relationship governing the discharge of a capacitor into a resistor. In practical circuits, the capacitor is sufficiently large to maintain some voltage across the resistor, and as the decreasing a.c. supply voltage falls below the voltage on the capacitor, the voltage across the rectifier diode decreases to zero and the current

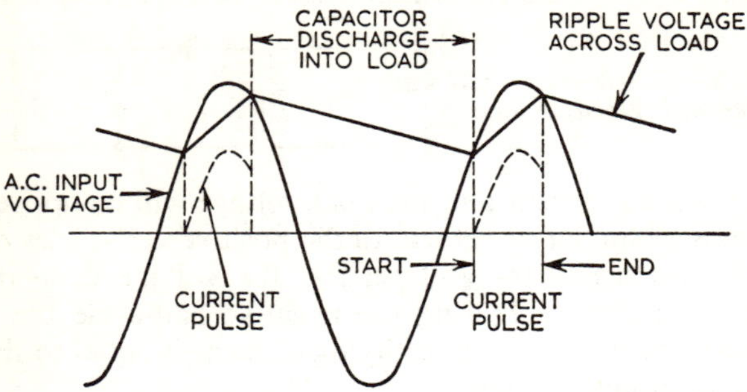

Fig. 5.—*Approximate waveforms of a half-wave capacitor input circuit.*

through the diode stops, generally just after the peak of the a.c. voltage wave is passed. Typical waveforms of the voltage and current in the circuit are shown in Fig. 5.

The fundamental difference in the mode of action of a circuit with a reservoir capacitor will now be clear. Without the capacitor the current flows through the diode for practically the whole positive half-cycle of the supply voltage, but with a reservoir capacitor the current passes in the form of a small pulse lasting for only a small fraction of the positive half-cycle. During the portion of the half-cycle when the diode is not conducting, the load current is supplied by the capacitor, the voltage falling exponentially until it is picked up again on the following half-cycle by the rising voltage across the transformer secondary winding. As the whole of the d.c. load power must be supplied by the transformer, the average value of the current pulse over the whole supply voltage cycle must equal the mean d.c. current, and thus the peak value of the current pulse must exceed the mean value by a large factor. In practical circuits of the half-wave type, the peak value of current pulse may be ten times the mean d.c. current in the load. Increase in the size of the reservoir capacitor will reduce the rate at which the voltage on the capacitor falls as it discharges into the load resistance, and will therefore decrease the time during which the supply voltage exceeds the voltage on the capacitor. For a given d.c. load current, the shorter current pulse must have a higher peak value. As the ripple voltage that exists across the reservoir capacitor is equal to the amount the capacitor voltage falls during the discharge period, an increase in the value of capacitor must decrease the ripple voltage. The process has been explained in some detail because it is fundamental to practically all the rectifier circuits commonly used.

In the half-wave circuit without a shunt capacitor across the load, the voltage across the diode on the non-conducting half-cycle is equal to the a.c. supply winding voltage $E_{a.c.}$, but when a capacitor is shunted across the load the diode will have to withstand an inverse voltage equal to the d.c. output voltage plus the a.c. supply voltage, the peak value of the inverse voltage being roughly $2\sqrt{2} \times E_{a.c.}$.

Single-phase Full-wave Rectifier

A circuit which is widely used in electronics is the single-phase full-wave circuit using a centre-tapped transformer: Fig. 6. This may be considered as two half-wave rectifiers

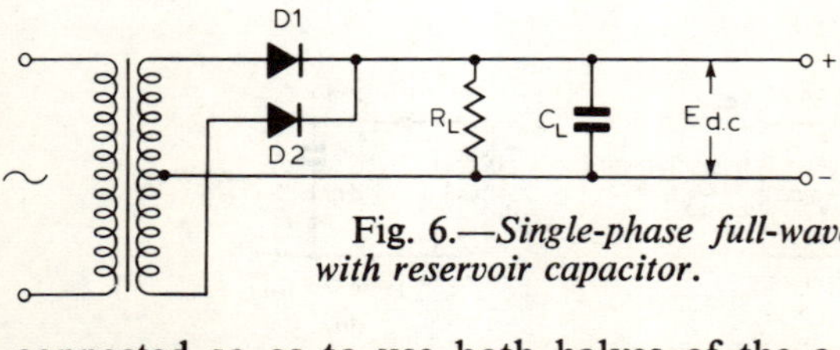

Fig. 6.—*Single-phase full-wave rectifier circuit with reservoir capacitor.*

connected so as to use both halves of the a.c. input. Each diode delivers current in turn during successive half-cycles; thus the two halves of the secondary winding and the two rectifier paths are used alternately. For low voltage appliances the two rectifiers may be combined in a single rectifier unit.

The pulsating d.c. output will follow the half-cycles and will thus be at twice the frequency of the a.c. input. The d.c. output into a resistive load will be twice that of the half-wave rectifier circuit, thus $E_{d.c.} = 0{\cdot}9E_{a.c.}$ (across each half of the transformer) and $I_{a.c.} = 0{\cdot}707I_{d.c.}$. On the reverse half-cycle the peak inverse voltage across each rectifier will rise to a value roughly equal to the d.c. voltage plus the peak a.c. voltage.

Full-wave Bridge Rectifier

The centre-tap of the transformer can be eliminated, and the input a.c. voltage and the peak inverse rating of the rectifiers

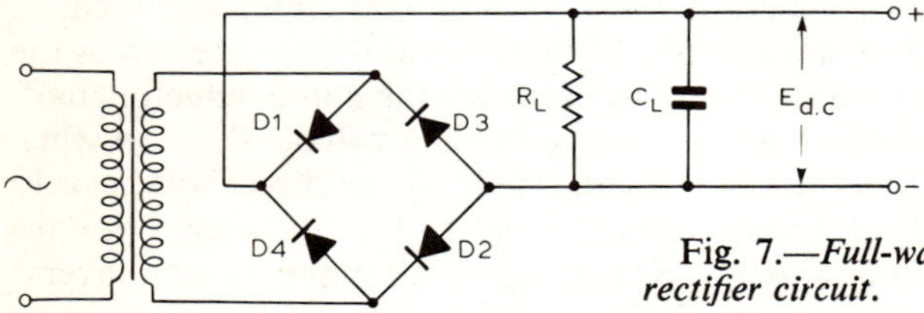

Fig. 7.—*Full-wave bridge rectifier circuit.*

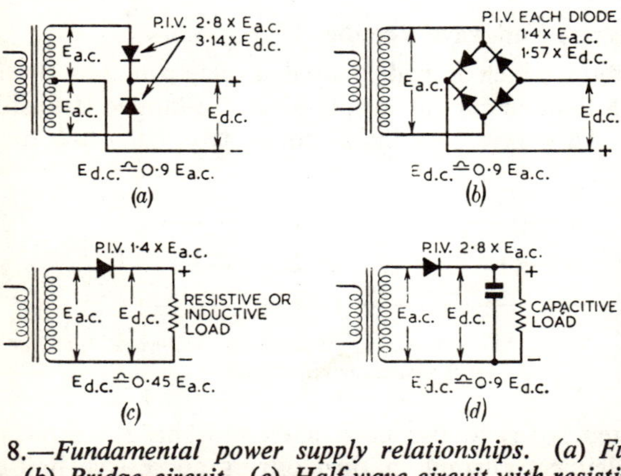

Fig. 8.—*Fundamental power supply relationships.* (*a*) *Full-wave circuit.* (*b*) *Bridge circuit.* (*c*) *Half-wave circuit with resistive load.* (*d*) *Half-wave circuit with capacitive load.*

halved, by the use of the full-wave bridge rectifier: the permissible output current, however, for a given transformer is halved, as the utilisation of the transformer winding is high. The output d.c. voltage is approximately equal to 0·9 of the a.c. voltage across the transformer secondary.

When the anode of D1 (Fig. 7) is positive, D1 and D2 in series conduct, current passing through the load. On the next half-cycle the anodes of D3, D4 will be positive and the current flows through them and through the load. Thus two valves are in series at all times, and the peak reverse voltage across any one is equal to the peak value of $E_{a.c.}$ voltage. The output waveform is similar to that of the single-phase full-wave rectifier.

Voltage-doubling Rectifier Circuits

Rectifiers may be connected so that they produce a considerably greater d.c. voltage output than the a.c. voltage input. Two voltage-doubler circuits are shown in Fig. 9 (a) and (b).

In the voltage-doubler full-wave circuit (Fig. 9 (a)), when point X is positive rectifier D1 will charge C1 to approximately

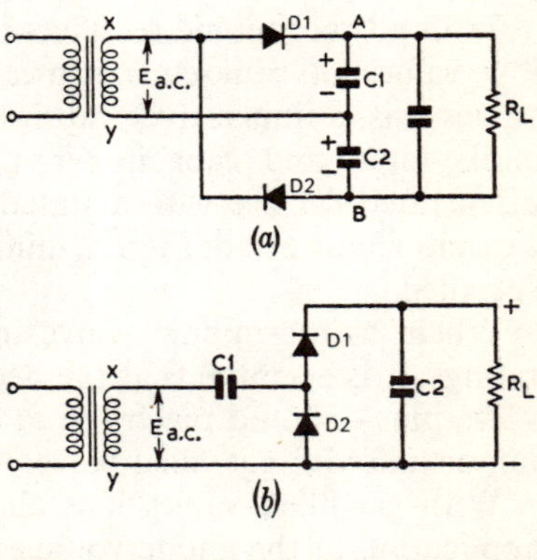

Fig. 9.—(*a*) *Full-wave voltage-doubler rectifier circuit.* (*b*) *Half-wave voltage-doubler rectifier circuit.*

the peak voltage of the a.c. input ($\sqrt{2}$ of the r.m.s. value), assuming little or no load. On the succeeding half-cycle, when Y is positive, D2 will charge C2 to a similar voltage, with the polarity indicated. There will thus exist, between points A and B, a d.c. voltage equal to $2\sqrt{2}$ of the a.c. input voltage. This assumes no load, and in practice with this type of circuit only small loads can be taken without poor voltage regulation.

The half-wave voltage-doubler circuit shown in Fig. 9 (b) is often preferred since it permits one side of the transformer winding to be earthed (and can thus also be used in transformer-less circuits).

The operation is as follows: On the half-cycle during which Y is positive, current flows into C1 through D2, leaving it charged to a voltage range $\sqrt{2}E_{a.c.}$ at the end of the half-cycle. On the succeeding half-cycle, the voltage applied to D1 will be the sum of the transformer output voltage and the voltage on C1. D1 will thus charge C2 to a peak voltage of approximately $2\sqrt{2}E_{a.c.}$. This system may be applied to produce greater voltage multiplications for high voltage, low current applications.

Thermionic Rectifier Precautions

With all thermionic rectifiers it is advisable to operate with low values of peak-current/mean-current and with gas-filled valves this is imperative: ratings may thus differ for use with choke-input and capacitor-input filters. Valve rectifiers are often rated for use with a stated maximum value of reservoir capacitor and in addition a minimum series resistance may be specified.

When a thermionic valve operates near its maximum ratings, it is essential that the heater voltage—measured at the valve pins—should reach the rated value. This is particularly important with gas-filled valves.

With gas-filled valves it is also good practice to delay the application of the anode voltage in order to allow the cathode to reach full operating temperature—this can be done by means of a delay switch.

RIPPLE REDUCTION

It has been noted that the output directly from the rectifier is in the form of pulsating d.c. Analysis of this output voltage would show it to comprise a d.c. component and an a.c. component whose frequencies are multiples of the a.c. supply frequency; the ripple harmonics contribute to the total magnitude and may usually be disregarded.

Table 1.—Ripple Frequencies and Magnitudes

Circuit	*Fundamental ripple frequency*	*Percentage peak ripple magnitude in terms of the d.c. component*
Single-phase half-wave . .	f	157
Single-phase, bi-phase full-wave	$2f$	68
Single-phase, full-wave bridge .	$2f$	68

In order to reduce the ripple voltage to negligible proportions the output is passed through a low frequency filter: these are usually combinations of capacitance and inductance or, where

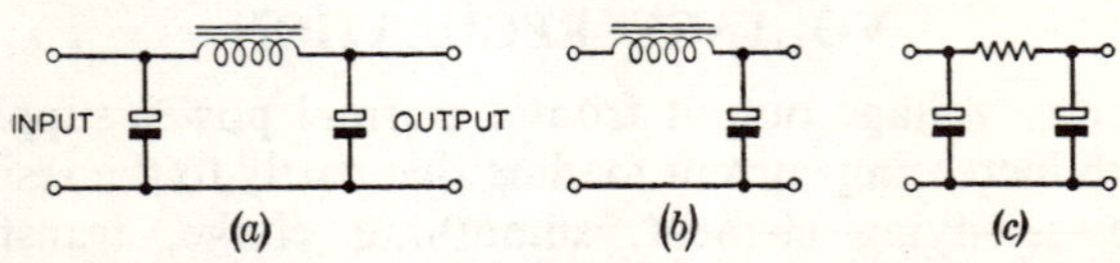

Fig. 10.—*Basic single-section filter circuits.* (*a*) *π-section capacitor input filter.* (*b*) *Choke-input filter.* (*c*) *Capacitance/resistance filter.*

output regulation and power losses are unimportant, capacitance and resistance. Basic single-section filters are shown in Fig. 10. Where a high degree of ripple reduction is required two or more filters might be used.

The action of the filter can be considered in the following simplified manner.

An inductance or choke offers impedance to any change in current and so tends to oppose the passage of pulsating current, whereas a capacitor blocks the passage of d.c. but offers only little impedance to the a.c. component. The a.c. component is thus opposed by the inductive (or resistive) element and short-circuited by the capacitors.

A filter with a capacitor as the first component (capacitor-input filter) provides a relatively high d.c. output due to the reservoir action of the input capacitor. The output voltage, however, drops appreciably as the load increases, and the rectifier is called upon to deliver high peak currents. With a choke input filter the voltage regulation is better, and there is less stress on the rectifier, but the maximum output voltage is reduced. With a capacitor input filter on higher-voltage supplies, the reservoir capacitor is normally limited in value to 16 μF with full-wave rectification and 32 μF with half-wave rectification, in order to limit peak rectifier current and ripple current through the capacitor. On lower-voltage supplies the reservoir may be some 2,000 μF.

Since inductive components are more expensive to produce than resistance, considerable use is made of resistance-capacitance filters for applications where the voltage drop and power loss in the resistor is less important.

VOLTAGE REGULATION

The d.c. voltage output from a normal power supply will fall with increasing current loading due partly to the resistance of the rectifying element, smoothing choke, transformer winding, etc. and partly to the changing relationship between the storage charge in the reservoir capacitor and load. If the value of the load and the input supply voltage remain steady during operation, the output will also remain steady. In some applications, however—for example where valves are operated in Class B or Class C or where switching sequences occur—this will not be the case. It may then be necessary to incorporate electronic stabilisation.

Voltage Regulator Tubes

Voltage regulator or stabiliser tubes are gas-filled diodes with unheated ("cold") cathodes, which are designed to give a stable burning voltage (the minimum voltage between anode and cathode required to *maintain* discharge) of particular value (for example 150 volts). For stabilising a load of about 20 mA, a voltage regulator tube may be connected as shown in Fig. 11. Since the ignition voltage required to start the discharge is higher than the burning voltage, the unregulated

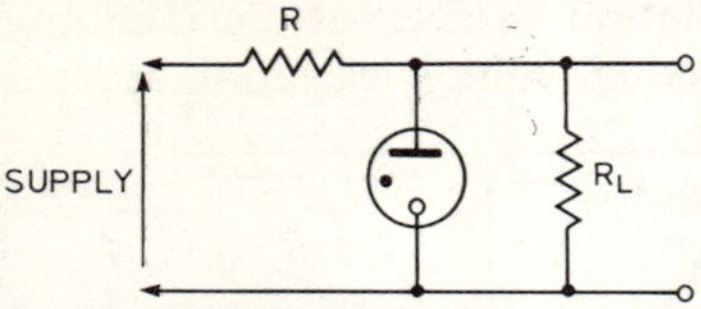

Fig. 11.—*Voltage regulator tube for stabilising a load of about 20 mA.*

supply must be of appreciably higher voltage than the output voltage required, and the value of R chosen so that the minimum ignition voltage is applied to the tube even under the most unfavourable conditions of high loads and low mains voltage.

Where it is required to stabilise a supply of more than a few milliamperes, a voltage regulator tube may be used to control a valve of variable impedance in series or parallel with the supply voltage. Voltage regulator and stabiliser tubes are further discussed in Section 14. The semiconductor equivalent of the voltage regulator tube is the zener diode (see Section 3).

Fig. 12 is a basic circuit of a series-regulated supply. The principle of operation is as follows: A rise in line voltage makes

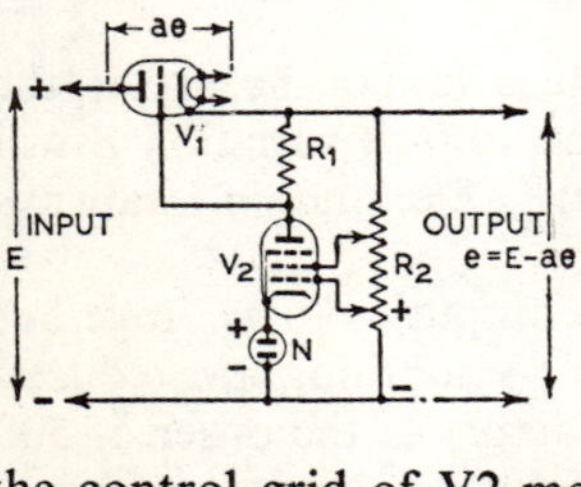

Fig. 12.—*Basic series d.c. thermionic voltage stabiliser circuit.*

the control grid of V2 more positive and causes the valve to pass a larger current. The increased voltage drop in R1 reduces the anode voltage and biases the grid of the regulating valve V1 more negatively. This action is augmented by the fact that the resistance of a neon lamp falls with rise of voltage. As a consequence the current increases to a greater extent than it would with an ohmic resistance of equal value. Provided R1 and R2 are suitably adjusted, the voltage change across valve V1 largely compensates for the rise of line voltage. If the line voltage falls, the reverse action occurs.

In practice 100 per cent regulation is never attained in any voltage regulator. If *e* is the residual voltage variation applied

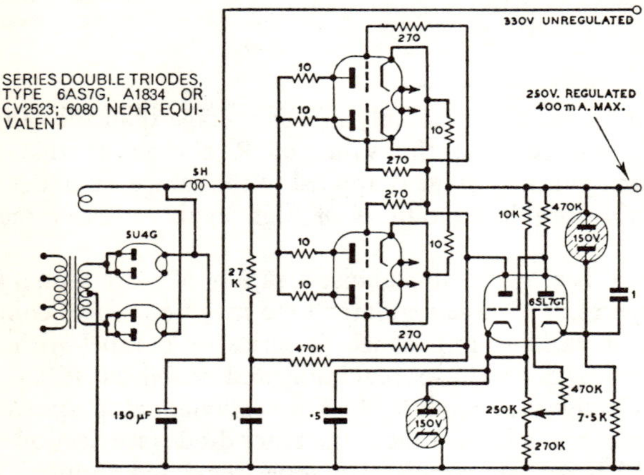

Fig. 13.—*Practical series-regulated power supply providing a stabilised* 250 *V output at up to* 400 *mA.*

to the control circuit and *a* is the voltage gain of the circuit, the correction voltage opposed to the line voltage variation *E* will be *ae*, so that $e = E - ae = E/(1+a) \simeq E/a$ since *a* is always much greater than unity.

Hence variations in the rectified output voltage, together with any residual ripple voltage after smoothing are reduced to $1/a$ of the original values. Gain factors of the order of 500 are commonly obtainable with well-designed circuits.

A representative design for a power supply providing a stabilised 250-volt output at up to 400 mA is shown in Fig. 13. In order to provide the necessary current-carrying capacity, four triode sections are connected in parallel to provide the variable series resistor element.

Voltage control can also be achieved by the use of power transistors in conjunction with a zener diode to provide the

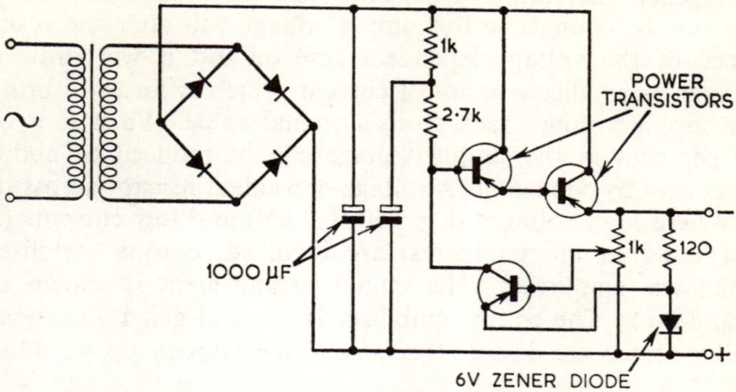

Fig. 14.—12-*V stabilised power supply using power transistors.*

reference voltage. Fig. 14 shows a typical low voltage regulated supply; this is a series type of regulator with feedback and a zener diode providing a 6 V reference level.

Voltage Stabilisation by Thermistors

The effect of temperature increases on a supply voltage may be reduced by the use of temperature-sensitive resistors (thermistors). The basic circuit arrangements are shown in

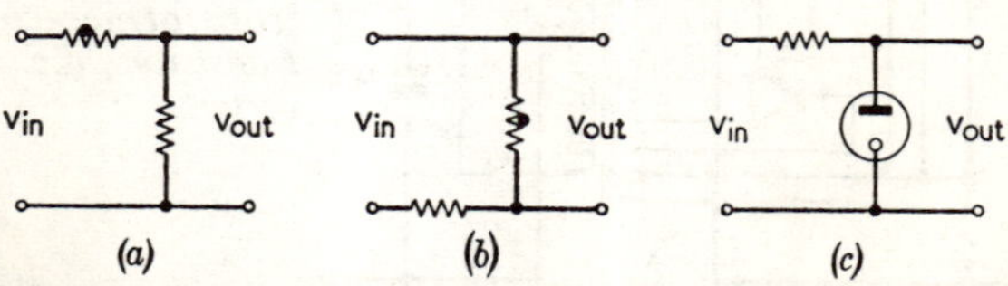

Fig. 15.—(*a*) *and* (*b*) *Use of a thermistor for voltage stabilisation against temperature increases.* (*c*) *The corona stabiliser.*

Fig. 15: (a) provides a more constant output voltage where the input voltage decreases with rise in temperature, and (b) a more constant output voltage where the input voltage increases with rise in temperature.

Voltage-dependent resistors are also available and may be used to stabilise circuits against variations in the supply voltage. If the voltage-dependent resistor is connected across the supply, changes in the supply voltage will alter the resistance of the voltage-dependent resistor and it will draw a greater or smaller amount of current, which in turn will bring the supply voltage back to its nominal value. Variations of 10 per cent in the supply voltage can be reduced to about 3 per cent by connecting a voltage-dependent resistor across it.

Where high voltages (e.g. 350 V–7 kV) and low currents (a few hundred microamperes) are involved, corona stabiliser tubes are employed. The circuit arrangement is shown in Fig. 15 (c). The corona stabiliser is a useful general-purpose voltage reference device. (See also Zener Diodes, pp. 42–43.)

D.C. CONVERTERS

Transistor d.c. converters provide an efficient means of converting d.c. to a.c. or to d.c. of a higher potential at powers up to a few hundred watts. Transistor d.c. converters can have an efficiency of the order of 85–90 per cent. The basis of

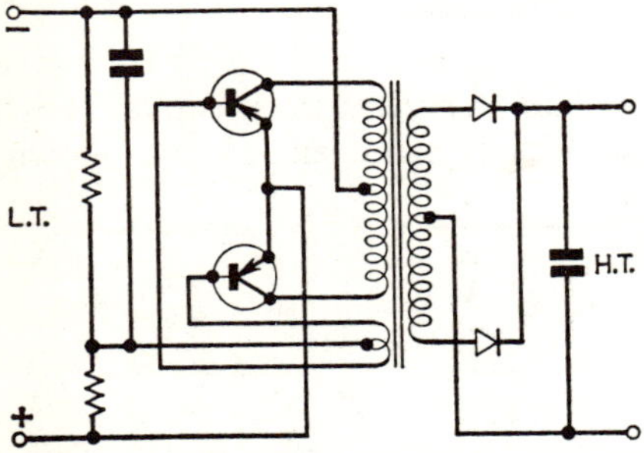

Fig. 16.—*Push–pull transformer-coupled transistor d.c. converter.*

conversion is that the d.c. input is "chopped" into a repetitive series of pulses which can be stepped up by normal transformer action. The transistors in a transistor d.c. converter provide this chopping, acting as on–off switches.

The most commonly used circuit arrangement is the push–pull transformer-coupled d.c. converter shown in Fig. 16. This will provide an output of over 100 watts from a 24-volt input. The two transistors conduct alternately, giving an a.c. output which is rectified to give a higher voltage d.c. output.

For higher powers, at present up to about 10 kW, silicon-controlled rectifiers may be used in inverter circuits: by careful design of the triggering circuits it is possible to obtain a near sinusoidal waveform.

SILICON POWER RECTIFIERS

Because of the low forward resistance of silicon rectifiers it is necessary to protect them against the surges that may occur when switching equipment on and off. In a single-phase

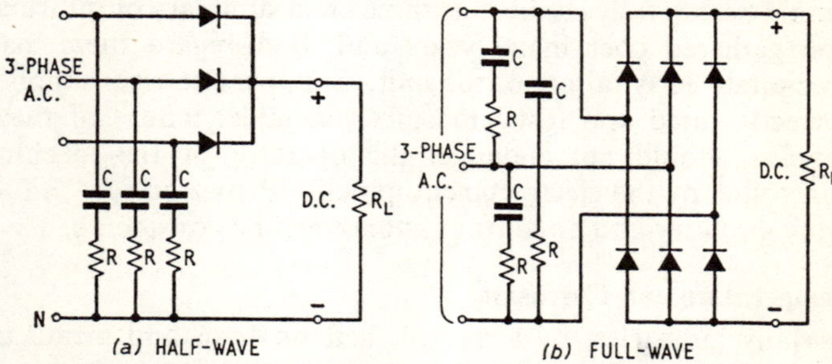

Fig. 17.—*Three-phase rectifier circuits with surge protection.*

circuit with smoothing, for example, there is the surge that occurs as the smoothing capacitor charges on switching on. Protection may be provided by means of a series surge-limiting resistor, generally of about 12–15 ohms. A by-pass capacitor may be used for protection against transient over voltages. In circuits fed from a transformer it is necessary to provide protection against the voltage surge that occurs when the transformer primary is switched off, and also surge due to oscillation when the primary is switched on. In the basic three-phase circuits shown in Fig. 17 capacitors C provide protection against the former and damping resistors R against the latter.

17

INSTALLATION AND SAFETY

INSTALLATION

Electronic equipment is now in use in nearly all forms of industry, and the location of such equipment relative to its environment is the first point to be considered. The following points should be observed whenever new equipment is to be installed. Instructions issued by the equipment manufacturers should be carefully studied, as these are a summary of information gathered over many years and to disregard them may precipitate early failure of the unit. Suitable electrical supplies correctly rated and fused must be available; failure of these supplies should not endanger the operator or the machine controlled by the electronic equipment. Provision for "fail to safe" circuitry and an alarm system must be considered.

Temperature and Corrosion

Many industries, by virtue of their methods and products, operate in conditions that are antipathetic to electronic equipment. High and low temperature extremes, corrosive atmospheres, and heavy vibration can all shorten the useful life of electronic units. Temperature limits specified for equipment should not be exceeded, and preferably a ten degree reserve on the maximum figure should be maintained to take care of extremes due to climatic changes. Transistors, having themselves a very low thermal inertia, are damaged very easily by short excursions outside the recommended maximum temperature range. While a catastrophic failure is unlikely, permanent damage will occur which will affect long-term accuracy and reliability. Low temperature extremes seldom promote early failures in electronic equipment, though they may have distinct effects on the accuracy of recorders and measuring instruments where the manufacturer has compensated the movement for a limited range of temperature only.

Corrosion, whether due to normal atmospheric pollution, or to high concentrations of acid or alkali in the air, is probably the greatest danger to industrial electronic equipment. The metals used in such equipment are primarily non-ferrous and may be vigorously attacked by such vapours. Miniaturisation and the use of printed circuits in modern equipment have reduced the amount of metal available in such things as coils and connections, making them more vulnerable to failure due to corrosion. Installation of units in areas which are known to produce corrosive vapours, i.e. chemical plants, foundries, plating shops, should be avoided, though it is realised that this is not always possible. Units which must be placed in such atmospheres should be fully sealed with foam rubber or polyurethane foam strip at all panel edges, and have their controls brought out through tight-fitting grommets. Switches should be of the "flame-proof" variety, having a flexible rubber moulding covering the dolly and its hinge pin. Units producing heat must be ventilated, and this should be done by forcing air into the bottom of the case and removing hot air at the top. The hot air should not be evacuated into the corrosive atmosphere as, when the equipment is shut off and the fan is stopped, cooling of the air within the case will draw corrosive vapour into the instrument case where it will remain until the fan is restarted. Units producing no heat may be completely sealed as described earlier, and under extreme conditions the outer cases may be pressurised, at a fraction of a pound per square inch, from a small gas cylinder. Nitrogen is commonly used for this purpose.

Mechanical Damage

Actual physical damage to electronic equipment usually results from incorrect siting of the unit when it is installed. Equipment brought into already established industry is usually placed in odd spaces not required by the existing machines or their operators, and mechanical damage quickly follows. Installation engineers should inform management that such arrangements are both costly and dangerous, and that a parallel consideration when purchasing new equipment is where such equipment shall be sited.

Man and Machine

As electronic units are introduced into established industry, the operators concerned should be introduced at an early stage to the proposed new methods. Electronic units badly operated will produce bad results, and electronics will get the blame. This in turn will reflect upon the installation and maintenance engineer.

Safety Requirements

All electronic installations should comply with I.E.E. Regulations for Electrical Equipment of Buildings, 14th edition 1966 and supplements, with regard to connections to the electricity supply and earthing. In addition to the above regulations, installations must be protected against the ingress of oil, water, and dust produced by local processes. Ventilation ducts should be protected to prevent the blocking off of free air flow by packing cases or similar large objects. Emergency "STOP" switches should be clearly labelled and placed within easy reach of the operator. Fire extinguishers provided near electronic equipment should be of a type recommended for use on electrical fires, i.e. dry powder "cartridges", CO_2 gas, or those containing a non-conducting liquid such as carbon-tetrachloride. Any extinguisher containing water or acid such as the common "soda-acid" type should be clearly labelled "NOT to be used on electrical fires".

High Voltage Units

Many industrial electronic units contain circuits producing voltages in excess of 1,000 volts, and special safety precautions must be observed when installing, commissioning and using such equipment. All units producing voltages in excess of

300 volts should be clearly labelled "DANGER HIGH VOLTAGE". Such units should have interlocks on doors and panels, ensuring that when the unit is opened all high voltage circuits are rendered inoperative. For test purposes the engineer can over-ride the interlock and operate the unit with the protective covers removed. In so doing he automatically accepts the responsibility for his own safety. Any repairs which require that the interlock be made inoperative must not be undertaken by one man; a second person must be available to render assistance in the event of accidental electric shock. All persons engaged in the installation, testing, and maintenance of high voltage equipment should make themselves familiar with the medical treatment for electric shock.

Optical Units

Units containing optical units such as edge-positioning equipment, smoke detectors, counting and batching equipment and guard operating equipment pose particular problems for the installation engineer. Changing sources of natural and artificial light throughout the year may seriously affect the operational stability of such units, and a complete study of the lighting conditions in the proposed working area should be made before installation is commenced. Smoke detectors, counting equipment, and similar units where a beam of light is focused on to a remote cell should be sited where steam, dust, or the operator's hands do not momentarily obscure the cell face, thereby giving false indications. Equipment installed to protect the operator from injury by machines, e.g. automatic operation of press or guillotine guards, should be fitted in "tamper-proof" boxes and invariably be of the "fail to safe" type.

Recording Apparatus

Many modern processes demand accurate monitoring and recording of variables such as temperature, pressure, flow, etc. This instrumentation must in many applications go on for 24 hours a day. Recorders, properly installed, are capable of performing such duty for many years with only minimal maintenance. Manufacturers provide detailed instructions for the installation and correct operation of their instruments, and these should be carefully studied. Recommended heights are specified which enable the operator to read the instrument with ease and facilitate changing the charts when necessary. Care should be taken to ensure that the instrument is truly level on its mountings: a few degrees tilt in the direction of the needle movement will on some instruments account for a gross non-linearity in the readings. Operating temperature recorders in areas where the ambient temperature is in excess of that recommended by the manufacturer may lead to consistently low readings being shown on the charts. Connections from recorders to thermocouples should always be made with the recommended compensating lead, and a careful check should be made to ensure that thermocouples installed in furnaces or ovens sense the average temperature within the heated volume, and are not influenced by local flame sources or hot-spots. Remote heads used to record flow or level should be sited away from heavily loaded mains cables, as many such heads are of the variable-reluctance type and tend to produce erratic or inaccurate electrical outputs when installed near strong alternating current fields.

MAINTENANCE

Electronic equipment once installed becomes an essential factor in the running of industrial plant. Failure of the electronics causes production to suffer, and the user is anxious that breakdowns shall be as few and of as short a duration as possible. The service engineer should ensure that equipment is maintained in first class order, and that faults occurring during the working day are speedily dealt with. The various systems of maintenance and how they can best be employed are dealt with below. The engineer on the spot must study the equipment in his charge and decide which of the systems will best serve. Factors of cost, speed, and availability of skilled labour will need to be considered.

Systems of Maintenance

Manufacturers' Service.—Most manufacturers of electronic equipment offer after-sales service. This varies from regular visits by a well-qualified and works-trained engineer, who is fully able to strip and completely service the unit, to a small and heavily overloaded service bay in a remote factory, dealing with specific units sent back for examination. Standards of service offered vary widely. The cost of calling out an engineer from the factory may be very high, while the adjustment he is required to do may only take a few minutes. Mobile engineers normally receive calls for assistance when they return to their base; this may mean one or two days before the faulty equipment is put right. While this system of service relieves the user of all responsibility relating to service of installed equipment, it is the least satisfactory where production is tied to the continuous and efficient working of the electronic units.

Exchange Units.—Where electronic units comprise many sections, e.g. power supplies, amplifiers, controllers, recorders, etc., and it is essential that the equipment shall be functional for long periods between maintenance shut-downs, the purchase of spare chassis may be considered a justifiable increase of the original purchase price. Not all equipment is conducive to this system of maintenance, as the use of quickly interchangeable chassis pre-supposes that connections are by plug and socket, and that no major setting-up procedure is required when a new chassis is brought into service. The faulty chassis may be returned to the manufacturer for repair or serviced by the local maintenance staff. Delays in servicing a faulty chassis reduce the reliability of the whole equipment, and too often this system of maintenance fails because the replacement chassis is itself found to be faulty when it is required.

Servicing on Site.—Most industrial electronic equipment is serviced in this way: by a small group of engineers employed particularly to do this job. The essentials are the proper test equipment and a well-trained, intelligent group of service engineers. The standard of maintenance will vary directly with these two factors. The engineer in charge will normally use some of the services described previously but will supplement these in large measure with his own knowledge and skill.

Formulating a Service Programme

When consideration is given to formulating a service programme on electronic equipment, it is essential that the user and the engineer should agree on the scope of the services envisaged and how these can best be fitted to the work programme of the equipment. Too often maintenance schedules break down because managements do not co-operate in allowing equipment to be out of operation for maintenance during working hours. It must be impressed on executives and managers that breakdowns due to lack of maintenance usually occur at times of peak production and may require hours or days to put right. Regular maintenance can be fitted into periods of low production and usually takes only an hour or so to complete.

Deciding on the Service System to be used.—After a thorough study of the equipment installed, the system of maintenance must be decided (see "Systems of Maintenance" above). The following points should also be considered:

(*a*) Recommendations on servicing given in the makers' handbook.

(*b*) The usage of the equipment, i.e. the number of hours per day the equipment must work.

(*c*) What units or components are heaviest loaded during normal working.

(*d*) What failures can be expected due to operators' setting-up or operating errors.

(*e*) What spares should be stocked, and what delivery time is quoted by the manufacturer for the supply of such spares.

From the above information a start can be made on listing the maintenance work and setting the period between the different tasks. Most work will divide quite well into weekly, monthly, three-monthly, six-monthly, and yearly tasks.

There will be exceptions, requiring attention every day, but these are met with very infrequently. A suggested schedule of weekly, monthly, and three-monthly tasks is given at the end of this section.

Service Records.—The tasks should be written out when the equipment is installed, and kept in the form of a wall chart in the service department. Alternatively details of the tasks should be shown on the backs of the service record cards. Careful records should be kept of all breakdowns, service, and fitting of new components, together with accurate costing for such work. This information is valuable when studying the reliability of units or components, and provides the engineer and management with a good indication of overall costing when equipment is due for replacement. Large file cards will enable all relevant information to be tabulated. They are easily stored and readily accessible for reference. Cards should not be removed from the file except to check records or to fill in the latest information. The card should never be given to the engineer when going out on a service call, though he should be encouraged to read the card before leaving the service department. His report of the fault, and his work, should be recorded on the card when the unit is again operational. It is usual to have at least two full sets of handbooks for equipment: one copy should be kept in a central file and should be considered a "master" copy; the second copy should be allowed to be freely available to maintenance personnel. With equipment that is repeated in a plant or organisation it may be invaluable to issue each service engineer with a full set of manuals. Modifications to any equipment must be entered on the circuit diagram or these diagrams soon become inaccurate and useless.

Spares.—Initially a small stock of spares should be held; these should include such things as transistors, valves, electrolytic capacitors, variable resistors and pilot lamps, plus charts, pens, and inks required for recorders. Lists of spares held for each unit should be kept up-to-date, and modified as usage determines which components are most likely to fail in service.

Methods of Servicing

For the servicing of any electronic unit to be successful, the engineer must be provided with adequate test gear. Certain items of gear are essential, e.g. universal test meter, oscilloscope. Other items should be selected to suit the installation concerned. Multi-purpose instruments can quite often be purchased with advantage where a high degree of accuracy is not always required, e.g. resistance, inductance, capacitance (R.L.C.) bridge, pulse/sine wave generator, F.M./A.M. signal generator. Maintenance of the test gear itself should not be overlooked, and instruments should be checked against a standard or returned to the manufacturer for overhaul and calibration at least once each year.

Fault Conditions

Faults on electronic apparatus fall roughly into three groups:

(*a*) A fault which completely stops the apparatus from working.

(*b*) A fault which causes the unit to perform badly, i.e. lack of sensitivity or accuracy.

(*c*) An intermittent fault which causes the unit to show symptoms of either (*a*) or (*b*) above, but only at random intervals. The unit may perform well at other times.

Of these three categories (*a*) is probably the easiest to locate and diagnose. Many of such faults are self-indicating, i.e. meters or recorders associated with the unit give positive indications that the fault lies in a particular circuit. Visual inspection will frequently locate faults in this category, the complete failure of a transistor, valve, capacitor or resistor being easily seen by the engineer. If a careful inspection reveals nothing, a check of all voltages within the unit will often show up the faulty circuit. It is unlikely that the fault will be due to ageing of components or maladjustment.

Bad performance (*b*) may show up when the unit is given a routine check, or the operator may express concern when the unit fails to give expected results in service. Again, meters or recorders may show what part of the circuit is failing, though this is by no means as certain as in (*a*). Logical circuit tracing (see later) by checking on all voltages and, where possible, currents will finally lead the engineer to the faulty component. This type of work is at times tedious and always time-consuming; however, if the details of the fault symptoms and the fault itself are entered on the service record card, a second fault of this kind can be quickly located and cleared.

Intermittent faults (*c*) are without doubt the worst problem that the service engineer must face. Simple causes are loose leads, plugs and sockets, and damaged switches, and these should always be checked first. When all the obvious points have been checked, and the fault recurs, an exchange chassis offers a solution in that the faulty unit can be returned to the service department and tested over a long period with meters and indicators connected to the individual circuits. In this way the fault causes little disruption of the instrumentation, and the engineer can study the faulty chassis under laboratory conditions. Printed circuits are known to produce baffling intermittent faults which can be traced to minute cracks in the thin copper strip near soldered connections. Flexing the chassis board on which the copper is deposited often reveals the fault by holding the crack open long enough for a measurement to be taken. Faults which only become apparent when the equipment has been running for some hours are commonly due to a breakdown within a component due to the increased temperature. Removing the faulty chassis and running it on the service bench quite often fails to produce the fault, as the operating temperature of the unit is much lower out of its case and in free air. A cardboard box (suitably lagged) placed over the unit while it is operating on the bench will produce conditions similar to the actual installation and accelerate the inception of fault conditions. After servicing, all units should be "soak tested" for some hours to ensure that no further adjustment or repair is required before re-installation.

Logical Fault Finding

Different types of equipment will require a different approach to fault finding; however, a generalised system is given here which, if applied to most faulty units, will greatly assist the engineer to locate the trouble.

(*a*) Check that no short circuit exists on the main supply leads a.c. and d.c. circuits, plug in the unit and allow it to warm up.

(*b*) Study the unit carefully; in particular observe all components for signs of overheating. Note all meter readings, closing of relays, and movement of servo-motors. Pay particular attention to the striking of neon-regulators, and to any component that is discoloured or is dripping wax. While transistor circuits running at lower powers than equivalent valve circuits are less demonstrative of faults, resistors should be checked for any signs of overheating. It cannot be overemphasised that careful observation of the faulty unit will reveal faults almost as often as checks carried out with complex test gear.

(*c*) Check all supply voltages starting with the a.c. input to the mains transformer and working through the d.c. supplies to all stages of the unit. Measure anode, screen grid, and cathode voltages on valves and emitter, base and collector voltages (and preferably currents) on transistor units.

(*d*) Check through the signal path by injecting suitable voltages and frequencies at the input terminals of each stage, starting with the output stage and working back. Observe the output signal of the final stage on an oscilloscope. Alternatively, where the input signal to a piece of equipment is "specialised", that input signal should be used; check the unit stage by stage using an oscilloscope and probe, working forwards from the input.

Information obtained from a series of tests similar to those shown above may not locate the faulty component, but it will specify the general circuitry in which the fault lies. Reference to the circuit diagram should now resolve the fault down to one or two components which can be checked individually.

Maintenance of Complex Semiconductor Equipment

Because of the different ageing and reliability characteristics of semiconductors, and the ease with which semiconductor devices can be inadvertently damaged during routine maintenance checks, a rather different approach towards electronic maintenance is developing.

This is based on the principle that if the equipment is functioning satisfactorily it is preferable to "leave it alone" rather than to attempt to check the individual semiconductors and components by means of voltage and current measurements. Furthermore, increasing amounts of the total circuitry may be represented by fully encapsulated modules, integrated circuits, etc., which are not intended to be repaired in the event of failure: it has been said that eventually we may arrive at the stage of "throw-away" electronics.

In place of routine maintenance, greater use is being made of automatic performance monitoring and failure alarms, combined in those cases where continuous service is regarded as essential with automatic switching to redundant equipment.

Reliability

Because of the expense that can result from a minor fault putting elaborate equipment out of action, considerable research into reliability, e.g. component reliability testing programmes, has been undertaken in recent years. Reliability testing backed by statistical analysis makes it possible to establish "mean time between failure" figures (the average number of hours during which an equipment should remain operative without failure under specified operational and environmental conditions). Given m.t.b.f. figures the time per annum during which equipment will be inoperative can be taken into account in installation design.

18

REFERENCE DATA

Power

The following formulæ apply to d.c. circuits. The power required to drive current through the resistance of a circuit is given by:

$$W = E \times I.$$

By consideration of Ohm's Law (see page 12) this may be stated also in the following forms:

$$W = I^2R$$

or

$$W = E^2/R$$

where W is the power in watts, E the potential difference in volts, I the current in amperes, and R the resistance in ohms.

Resistors

Resistors in Series

$$R\text{ (total)} = R_1 + R_2 + R_3 \ldots.$$

Thus the total resistance of a chain of resistors in series is the sum of the individual resistors which form the chain.

EXAMPLE: *What is the effective resistance of a* 33*k resistor and a* 0·25*M resistor connected in series?*

Note that all values must be changed into a common unit before addition.

$$R = R_1 + R_2 = 33\text{k} + 0{\cdot}25\text{M} = 33{,}000 + 250{,}000$$
$$= \underline{283{,}000 \text{ ohms.}}$$

Resistors in Parallel

$$\frac{1}{R} = \frac{1}{R_1} + \frac{1}{R_2} + \frac{1}{R_3} \ldots.$$

Where there are only two resistors, this formula may be stated:

$$R = \frac{R_1 \times R_2}{R_1 + R_2}.$$

N.B.: The effective resistance of a number of resistors in parallel must always be less than that of the resistor with the lowest value.

EXAMPLES: (1) *What is the effective resistance of* 2,000-*ohm*, 5,000-*ohm and* 10,000-*ohm resistors connected in parallel?*

$$\frac{1}{R} = \frac{1}{2{,}000} + \frac{1}{5{,}000} + \frac{1}{10{,}000} = \frac{5+2+1}{10{,}000},$$

$$\therefore\ R = \underline{1{,}250 \text{ ohms.}}$$

(2) *What value resistor should be connected in parallel across a* 60,000-*ohm resistor in order to reduce the effective circuit resistance to* 20,000 *ohms?*

$$20{,}000 = \frac{60{,}000 \times R}{60{,}000 + R},$$

$$\therefore\ 20{,}000\,(60{,}000 + R) = 60{,}000R,$$

$$\therefore\ R + 60{,}000 = 3R,$$

$$\therefore\ R = \underline{30{,}000 \text{ ohms.}}$$

Potential Divider

Where two resistors are connected across a supply (see Fig. 1), the voltage across one, in the absence of a load, is derived from the following formula:

$$e = E \times \frac{R_1}{R_1 + R_2}.$$

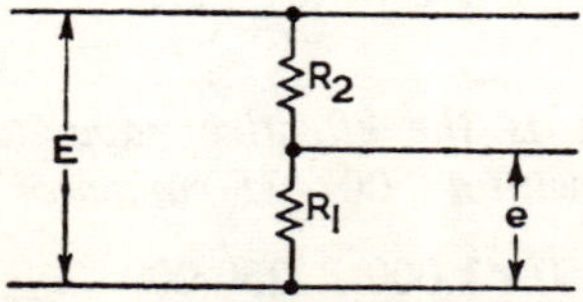

Fig. 1.—*Potential divider.*

Where a load is drawn from a potential divider, the differing currents in each section of the divider must be taken into account.

EXAMPLE: *A potential divider is to be connected across a* 200-*volt h.t. line so as to provide an output of* 100 *volts at* 5 *mA with a standing current of* 4 *mA, see Fig.* 2.

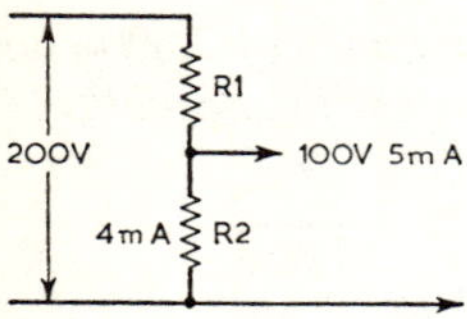

Fig. 2.—*Potential divider—example.*

The current flowing through R_1 will be $5+4 = 9$ mA. Since R_1 must provide a voltage drop of 100 volts, then, from Ohm's Law,

$$R_1 = \frac{E}{I} = \frac{100}{9} \times 1{,}000 \simeq 11{,}000 \text{ ohms}$$

and $$R_2 = \frac{100}{4} \times 1{,}000 = \underline{25{,}000 \text{ ohms.}}$$

Capacitors

Capacitors in Series

$$\frac{1}{C} = \frac{1}{C_1} + \frac{1}{C_2} + \frac{1}{C_3} \cdots .$$

Where there are only two capacitors, this formula may be stated:

$$C = \frac{C_1 \times C_2}{C_1 + C_2}.$$

EXAMPLE: *What is the effective capacitance of a* 250-*pF capacitor in series with a* 0·001-*μF capacitor?*

$$C = \frac{250 \times 1{,}000}{250 + 1{,}000} = \frac{250{,}000}{1{,}250} = \underline{200 \text{ pF.}}$$

Capacitors in Parallel

$$C = C_1 + C_2 + C_3 \ldots.$$

EXAMPLE: *What is the effective capacitance of a* 0·0002-μF *capacitor in parallel with a* 50-*pF capacitor?*

$$C = 200 + 50 = \underline{250 \text{ pF}}.$$

Inductors

Inductances in Series

$$L = L_1 + L_2 + L_3 \ldots.$$

Inductances in Parallel

$$\frac{1}{L} = \frac{1}{L_1} + \frac{1}{L_2} + \frac{1}{L_3} \ldots.$$

Resonant Circuits

The reactance of an inductor (X_L) to an a.c. signal is equal to $2\pi fL$, where f is the frequency in hertz, L the inductance in henrys and π is approximately 3·14. The reactance of a capacitor (X_C) to an a.c. signal equals $1/2\pi fC$, C being in farads and the other units as before. A more practical formula is:

$$X_C = \frac{1{,}000{,}000}{2\pi fC}$$

where X_C is the capacitive reactance in ohms, f the frequency in hertz, C the capacitance in microfarads and π is approximately 3·14.

EXAMPLE: *What is the reactance of a* 0·1-μF *capacitor at* 5,000 *Hz?*

$$X_C = \frac{1{,}000{,}000}{2 \times 3{\cdot}14 \times 5{,}000 \times 0{\cdot}1} = \frac{200}{0{\cdot}628} = \underline{318 \text{ ohms.}}$$

The values shown in Table 1 are typical of those commonly encountered.

Capacitive reactance decreases with increase in frequency, inductive reactance increases with increase in frequency. Where inductance and capacitance are combined the resultant reactance of the combination is the difference between them, i.e. $X = X_L - X_C$. At one particular frequency, the resonant frequency, the inductive and capacitive reactances are equal

Table 1

C	X_C *(ohms)*				
	50 *Hz*	100 *Hz*	1,000 *Hz*	5,000 *Hz*	15,000 *Hz*
100 pF	31,800,000	15,900,000	1,590,000	318,000	106,000
500 pF	6,370,000	3,180,000	318,000	63,700	21,200
0·001 μF	3,180,000	1,590,000	159,000	31,800	10,600
0·005 μF	637,000	318,000	31,800	6,370	2,120
0·01 μF	318,000	159,000	15,900	3,180	1,060
0·1 μF	31,800	15,900	1,590	318	106
1·0 μF	3,180	1,590	159	31·8	10·6

C	X_C *(ohms)*				
	500 *kHz*	1 *MHz*	5 *MHz*	10 *MHz*	50 *MHz*
10 pF	31,800	15,900	3,180	1,590	318
50 pF	6,370	3,180	637	318	63·7
100 pF	3,180	1,590	318	159	31·8
500 pF	637	318	63·7	31·8	6·37
0·001 μF	318	159	31·8	15·9	3·18
0·005 μF	63·7	31·8	6·37	3·18	0·637

and cancel out, so that there is no net reactance and the only impedance is the d.c. resistance of the components. In a resonant circuit in which the inductive and capacitive components are connected in series, impedance is minimum and current flow maximum at the resonant frequency: such a circuit can be used to select signals at the resonant frequency and reject those at other frequencies or to short-circuit unwanted signals at the resonant frequency. Where the inductive and capacitive components are connected in parallel, impedance is at the maximum and current flow minimum at the resonant frequency: this type of circuit can thus be used to filter out signals at the resonant frequency. A parallel resonant circuit can also be used to provide frequency selection since as at resonance it is effectively a pure resistance it can be used as a frequency selective load for a voltage amplifier. The effective resistance R_E at resonance of a parallel resonant circuit equals L/Cr, where r is the d.c. resistance in ohms of the circuit. The resonant frequency of a circuit consisting of an inductance and a capacitance is given by the formula $1/2\pi\sqrt{LC}$. A more practical formula is:

$$f = \frac{1{,}000{,}000}{2\pi \times \sqrt{(L \times C)}}$$

$$\simeq \frac{159{,}000}{\sqrt{(L \times C)}}$$

where f is the resonant frequency in kilohertz, L is the inductance in microhenrys and C is the capacitance in picofarads.

The ratio of the reactance to the resistance of a coil, capacitor or resonant circuit is termed its Q factor, and since for a frequency selective circuit the resistance is usually unwanted, Q is regarded as a measure of "goodness". The Q of a capacitor being very much higher than that of an inductor, the Q of a parallel resonant circuit can generally for practical purposes be regarded as that of the inductor.

The higher the Q (ratio of reactance to resistance), the sharper is the difference in response of a resonant circuit to signals at the resonant frequency from those at nearby frequencies. Maximum selectivity may not, however, be the circuit designer's aim. Where wide bandwidth in addition to frequency selective characteristics is required, additional resistance may be added to the circuit to reduce its Q—see Fig. 3 (a).

If two parallel circuits are mounted close together so that they are magnetically coupled, their interaction tends to mistune them from resonance, so that adjustment of the amount of coupling can be used to produce a wideband response, see Fig. 3 (b).

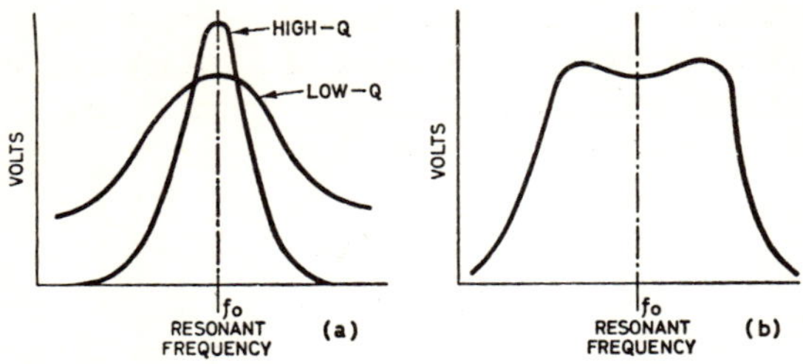

Fig. 3.—(*a*) *Gain-frequency characteristics of high- and low-Q resonant circuits.* (*b*) *Bandpass characteristic obtained by coupling between two resonant circuits.*

Alternating Current and Voltage

The instantaneous values of voltage (e) and current (i) are given by $e = E_{\text{peak}} \sin \omega t$ and $i = I_{\text{peak}} \sin \omega t$.

Root mean square values: $E_{\text{r.m.s.}} = 0{\cdot}707 E_{\text{peak}}$,

$I_{\text{r.m.s.}} = 0{\cdot}707 I_{\text{peak}}$.

Peak values: $E_{\text{peak}} = 1{\cdot}414 E_{\text{r.m.s.}}$, $I_{\text{peak}} = 1{\cdot}414 I_{\text{r.m.s.}}$

Note that peak-to-peak value is $2\sqrt{2}$ (i.e. 2·828) times $E_{\text{r.m.s.}}$.

Current in circuits containing inductance (L, henrys), capacitance (C, farads) and resistance (R, ohms):

$$I = \frac{E}{\sqrt[2]{\left\{R^2 + \left(\omega L - \frac{1}{\omega C}\right)^2\right\}}}.$$

Impedance, $Z = \sqrt{\{R^2 + (X_L - X_C)^2\}}$,

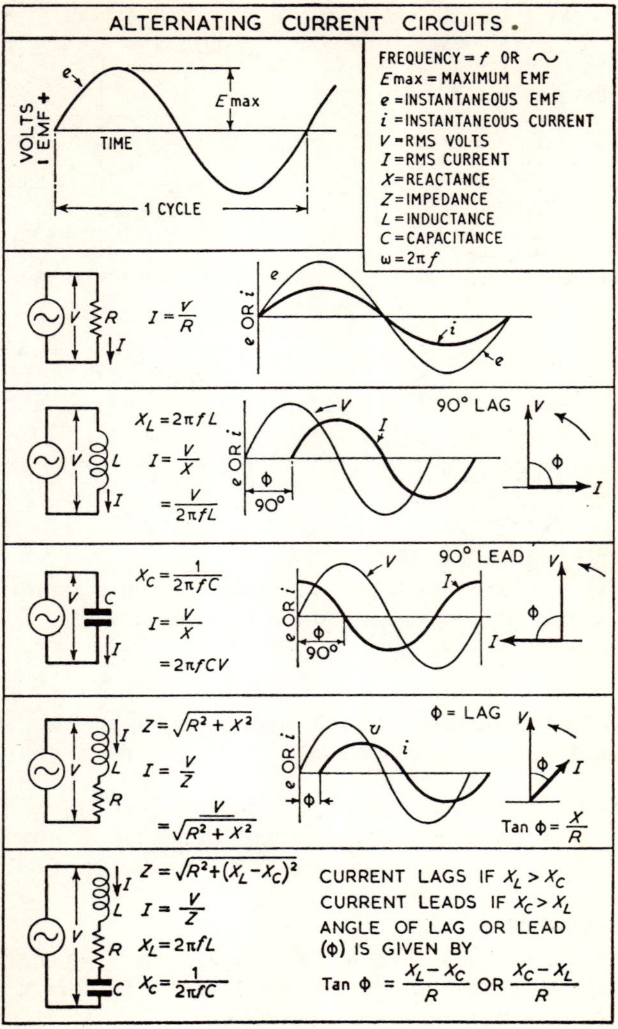
ALTERNATING CURRENT CIRCUITS
FREQUENCY = f OR ~
Emax = MAXIMUM EMF
e = INSTANTANEOUS EMF
i = INSTANTANEOUS CURRENT
V = RMS VOLTS
I = RMS CURRENT
X = REACTANCE
Z = IMPEDANCE
L = INDUCTANCE
C = CAPACITANCE
ω = 2πf
VOLTS + EMF −
TIME
E max
1 CYCLE
I = V/R
e OR i
X_L = 2πfL
I = V/X = V/2πfL
90° LAG
Φ
90°
X_C = 1/2πfC
I = V/X = 2πfCV
90° LEAD
Z = √(R² + X²)
I = V/Z = V/√(R² + X²)
Φ = LAG
Tan Φ = X/R
Z = √(R² + (X_L − X_C)²)
I = V/Z
X_L = 2πfL
X_C = 1/2πfC
CURRENT LAGS IF X_L > X_C
CURRENT LEADS IF X_C > X_L
ANGLE OF LAG OR LEAD (Φ) IS GIVEN BY
Tan Φ = (X_L − X_C)/R OR (X_C − X_L)/R

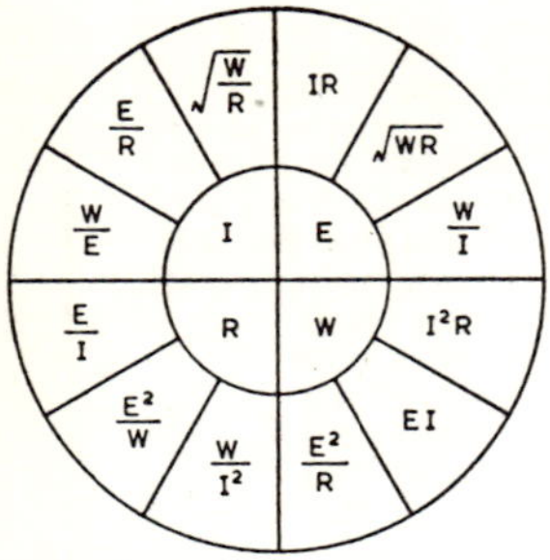

OHM'S LAW FOR D.C. CIRCUITS

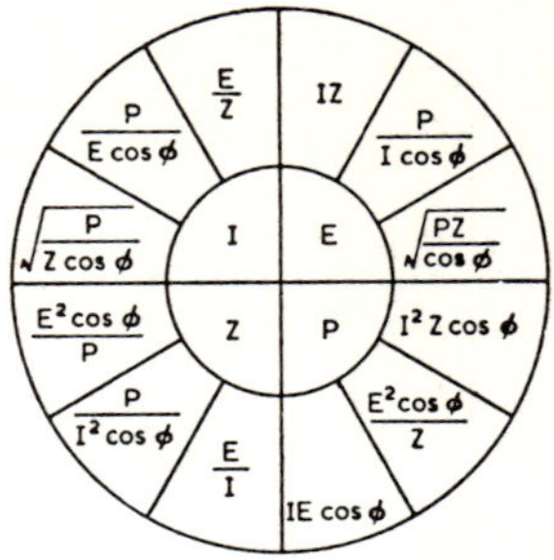

OHM'S LAW FOR A.C. CIRCUITS

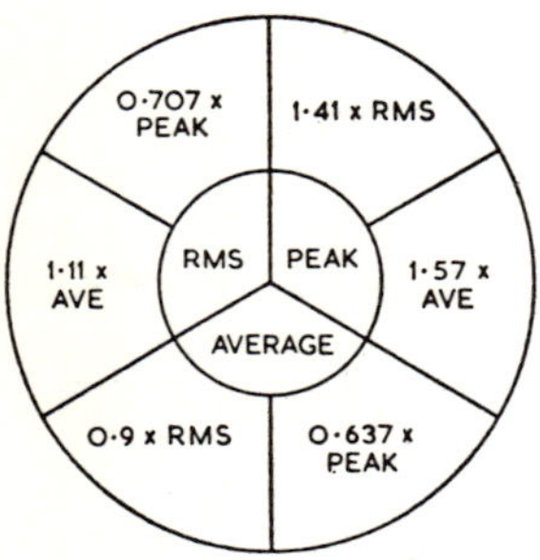

A.C. VOLTAGE RELATIONSHIPS

W WATTS E VOLTS I AMPS R OHMS
Z IMPEDANCE P TRUE POWER
φ PHASE ANGLE

Fig. 4.

X_L being inductive reactance and X_C capacitive reactance. Phase difference (series circuits),

$$\tan\phi = \frac{X_L - X_C}{R} = \frac{\text{Reactance}}{\text{Resistance}}.$$

Time Constant

If a capacitor (C, Fig. 5 (a)) is charged by a potential V_0 through a resistor R, the charge V across C will in time $t = R \times C$, where t is in seconds, R ohms and C farads, reach approximately $\frac{2}{3}$ of V_0. V never reaches the full potential V_0

because of the drop across R. Charging is, as shown at (b), exponential. The discharge of C through R similarly follows an exponential curve which is the mirror image of the charging curve.

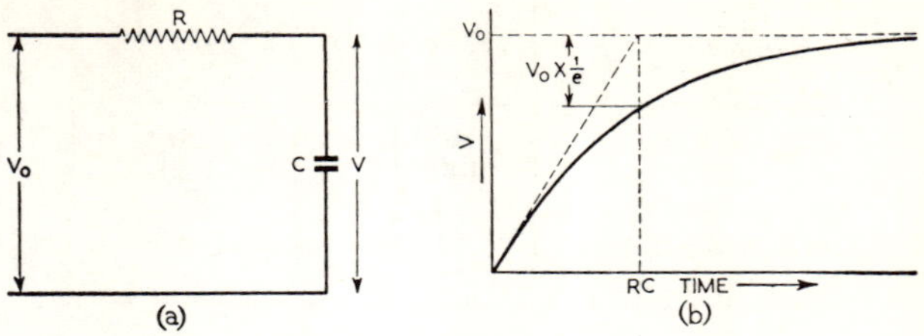

Fig. 5.—*Charging a capacitor through a resistor.*

In the case of an inductor connected to a potential via a series resistor, the time constant is the time taken for the current through the inductor to reach approximately $\frac{2}{3}$ of its maximum value, and is given by $t = L/R$, where t is again in seconds, L in henrys and R is the circuit resistance in ohms.

Attenuation or Gain

Many amplifier gain controls are calibrated in decibels (dB). The decibel is one-tenth of a Bel, which is numerically the logarithm (to base 10) of the power ratio; that is, if the power gain is 10^x, then the gain in Bels $= x = 10x$ dB. The Bel is the logarithm of a ratio and is not an absolute unit, and this ratio is a ratio of power and not voltage or current. The Decibel may be used for the comparison of voltages if the impedance is kept constant as in these circumstances power is proportional to (voltage)^2 and hence

$$\text{Gain (dB)} = 10 \log_{10} \frac{(V_{out})^2}{(V_{in})^2} = 20 \log_{10} \frac{V_{out}}{V_{in}}.$$

Some examples are given in Table 2. With this system the gains and attenuations of the various parts of an amplifying system can be added and subtracted instead of multiplied and divided.

Table 2

Voltage ratio $\frac{V_{out}}{V_{in}}$	*Power ratio*	dB
100 : 1	10^4 : 1	+40
10 : 1	100 : 1	+20
2 : 1	4 : 1	+ 6
1 : 100	1 : 10^4	−40

Noise

Noise arises in electronic equipment due to the random movement of electrons in conductors and the various components in the equipment. The noise factor of an equipment is defined as the ratio of the signal/noise ratio at the output to the signal/noise ratio at the input, and is generally expressed in dB. Since noise is often distributed over the frequency spectrum, reducing equipment bandwidth improves its noise factor. Noise contributed by an electronic device or piece of equipment is sometimes specified in terms of noise temperature: this is the noise it contributes compared with that which would be contributed by an equivalent pure resistance at room temperature (290°K); for example, a device contributing twice the noise of an equivalent pure resistance would have a noise temperature of 580°K.

Mark/Space Ratio

In pulse terminology the interval between pulses is commonly referred to as the space and the duration of the pulse the mark, giving rise to the expression mark/space ratio.

Frequency and Wavelength

The velocity at which electromagnetic waves travel in a given medium is constant. In space the velocity, *c*, is equal to that of light, i.e. approximately $2{\cdot}998 \times 10^8$ metres per second. For practical purposes c may be taken as 3×10^8 m/s. Frequency f in terms of velocity and wavelength λ in metres is given by $f = c/\lambda$. Thus

$$f(\text{Hz}) = 3 \times 10^8/\lambda = 300{,}000{,}000/\lambda,$$
$$f(\text{kHz}) = 3 \times 10^5/\lambda = 300{,}000/\lambda,$$
$$f(\text{MHz}) = 300/\lambda.$$

For frequency/wavelength conversion useful datum points are:

f(MHz)	1	3	10	30	100	300	1,000	3,000
λ(metres)	300	100	30	10	3	1	0·3	0·1

The velocity of sound (pressure waves) in air, v, is approximately 340 m/s. With wavelength λ given by v/f, the wavelength of, for example, a 500 Hz note is 340/500 or 670 cm approximately.

Testing Junction Transistors Quickly

First-line maintenance on equipment frequently leads to the problem of deciding whether a particular transistor has failed. In such circumstances transistor testers and equipment for obtaining transistor characteristics are not always convenient or at hand. However a multimeter usually is, and if the transistor can be removed from the circuit the following procedure usually allows a "go/no-go" decision to be made for a particular device on a fairly reliable basis. This technique is in no way intended to replace precise measurement of transistor parameters and characteristic curves on the proper instruments.

Identifying terminal configuration and transistor type (npn or pnp)

In many cases it will already be known, or will be apparent, which terminal is which (e.g. TO-3 package power transistors—casing is collector). If nothing is known on the configuration proceed as follows.

Starting with the $\Omega \div 100$ range, choose the lowest resistance range which yields a clear difference in resistance, with polarity reversal, between pairs of terminals. Using too high a range may damage the device and/or give misleading results. (Remember that on resistance ranges the *red* AVO terminal provides the *negative* connection and vice versa.)

Find the pair of terminals between which the resistance is *not low* (relatively speaking) to *either* polarity; the *remaining* terminal is the *base* (Fig. 6). From diode principles determine

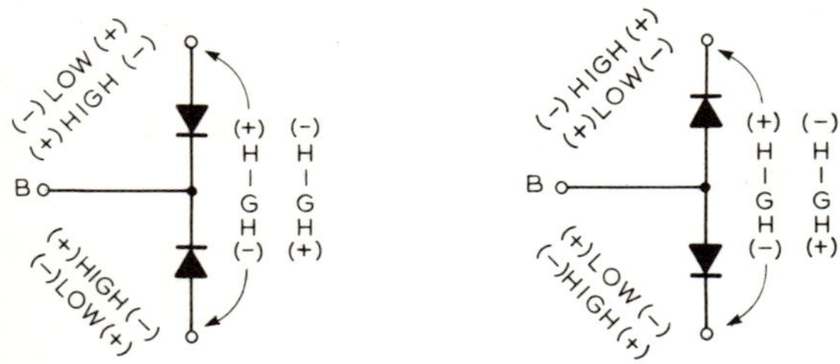

Fig. 6.—*Determining which terminal of a transistor is the base.*

whether the base is n-type or p-type and hence the transistor pnp or npn: if the junctions are *low* resistance with a *positive* connection to the base, the base must be p-type and the transistor npn, and vice versa (Fig. 7).

Fig. 7.—*Determining whether the base is n-type or p-type material.*

Distinguishing collector from emitter

Connect the meter across the pair of terminals, note the resistance and repeat with reversed polarity. The collector-emitter resistance is *highest* when the polarity of connection is that for "normal" operation, i.e. negative to the collector for pnp, positive to the collector for npn. Since the type has already been determined this test identifies the collector (and by elimination the emitter). In Fig. 8, the higher resistance reading for (a) shows that this is the "correct" supply polarity; since transistor type is known (say npn in this case), terminal 1 is the collector because an npn takes a positive collector rail.

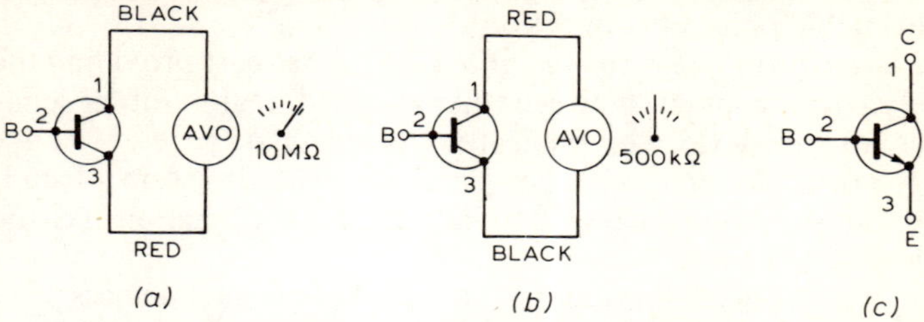

Fig. 8.—*Distinguishing between collector and emitter terminals.*

Leave the AVO connected for "normal" supply polarity between collector and emitter for the particular device.

Testing for transistor action

Assuming previous tests have indicated two "good" junctions, with a finger lightly touch (but do not short together) the collector and base leads simultaneously. A "good" transistor will yield a marked deflection of the meter (reduction of collector-emitter resistance).

If this test is carried out with collector and emitter incorrectly identified a deflection is usually obtained but the device exhibits considerably less "gain". Be wary of relying too much on this test when applied to large power transistors!

GENERAL SEMICONDUCTOR DEVICE TERMS*

General terms

Terminal (*of a semiconductor device*). A specified externally available point of connection.

Electrode (*of a semiconductor device*). That part providing the electrical contact between the specified region of a semiconductor device and the lead to its terminal.

Forward direction (*of a pn junction*). The direction of continuous (direct) current flow in which a pn junction has the lowest resistance.

Note. This definition may not apply to tunnel devices.

Reverse direction (*of a pn junction*). The direction of continuous (direct) current flow in which a pn junction has the higher resistance.

Note. This definition may not apply to tunnel devices.

Physical terms

Semiconductor. A material with resistivity usually in the range between metals and insulators, in which the electrical charge carrier concentration increases with increasing temperature over some temperature range.

Extrinsic semiconductor. A semiconductor with charge carrier concentration dependent upon impurities or other imperfections.

N-type semiconductor. Extrinsic semiconductor in which the conduction electron density exceeds the mobile hole density.

P-type semiconductor. Extrinsic semiconductor in which the mobile hole density exceeds the conduction electron density.

I-type (*intrinsic*) *semiconductor*. Nearly pure and ideal semiconductor in which the electron and hole densities are nearly equal under conditions of thermal equilibrium.

Junction. A region of transition between semiconducting regions of different electrical properties.

* The definitions of terms and symbols on pages 328–341 are reproduced by courtesy of Tektronix U.K. Ltd.

pn Junction. A junction between p and n type semiconductor material.
Alloyed junction. A junction formed by alloying one or more materials to a semiconductor crystal.
Diffused junction. A junction formed by the diffusion of an impurity within a semiconductor crystal.
Grown junction. A junction produced during the growth of a semiconductor crystal from a melt.
Charge carrier (*abbreviation: carrier*). In a semiconductor, a mobile (free) conduction electron or mobile hole.
Majority carrier (*in a semiconductor region*). The type of carrier constituting more than half of the total charge carrier concentration.
Minority carrier (*in a semiconductor region*). The type of carrier constituting less than half of the total charge carrier concentration.
Depletion layer. A region in which the mobile charge carrier density is insufficient to neutralise the net fixed charge density of donors and acceptors.
Breakdown (*of a reverse-biased pn junction*). A phenomenon, the initiation of which is observed as a transition from a state of high dynamic resistance to a state of substantially lower dynamic resistance for increasing magnitude of reverse current.
Avalanche breakdown (*of a semiconductor pn junction*). A breakdown that is caused by the cumulative multiplication of free charge carriers in a semiconductor under the action of a strong electric field which causes some free carriers to gain enough energy to liberate new hole-electron pairs by ionisation.
Avalanche voltage. The applied voltage at which avalanche breakdown occurs.
Thermal breakdown (*of a semiconductor pn junction*). A breakdown that is caused by the generation of free charge carriers owing to the cumulative interaction between increasing power dissipation and increasing junction temperature.

Note. This effect is also known as thermal runaway.

Zener breakdown (*of a semiconductor pn junction*). A breakdown caused by the transition of electrons from the valence band to the conduction band due to tunnel action under the influence of a strong electric field.

Zener voltage. The applied voltage at which zener breakdown occurs.

Tunnel effect. The piercing of a potential hill by a carrier, which would be impossible according to classical mechanics, but the probability of which is not zero according to wave mechanics, if the width of the hill is small enough. The wave associated with the carrier is almost totally reflected on the first slope, but a small fraction crosses the hill.

Tunnel action (in a pn junction). A process whereby conduction occurs through the potential barrier due to the tunnel effect and in which electrons pass in either direction between the conduction band in the n-region and the valence band in the p-region.

Note. Tunnel action, unlike the diffusion of charge carriers, involves electrons only and for all practical purposes the transit time is negligible.

Photo-electric effect. Interaction between radiation and matter resulting in the absorption of photons and the consequent generation of mobile charge carriers.

Photovoltaic effect. A photo-electric effect wherein an electromotive force is generated.

Types of device

Semiconductor device. A device whose essential characteristics are due to the flow of charge carriers within a semiconductor.

Semiconductor diode. A two-terminal semiconductor device having an asymmetrical voltage-current characteristic.

Note. Unless otherwise qualified, this term usually means a device with the voltage-current characteristic typical of a single pn junction.

Voltage reference diode. A diode which develops across its terminals a reference voltage of specified accuracy, when biased to operate within a specified current range.

Voltage regular diode. A diode which develops across its terminals an essentially constant voltage throughout a specified current range.

Semiconductor rectifier diode. A semiconductor diode designed for rectification and including its associated mounting and cooling attachments if integral with it.
Tunnel diode. A diode having a pn junction in which tunnel action occurs giving rise to negative differential conductance in a certain range of the forward direction of the current-voltage characteristic.
Transistor. A semiconductor device capable of providing power amplification and having three or more terminals.

Note. Other names may be used to describe certain special types of semiconductor device covered by this definition.

Photoconductive cell. A device in which the photoconductive effect is utilised.
Photovoltaic cell. A device in which the photovoltaic effect is utilised.
Photo-diode. A diode in which the photoelectric effect is utilised.
Photo-transistor. A transistor in which the photoelectric effect is utilised.

Terms related to ratings and characteristics

Reverse voltage. The voltage across a junction or a diode when biased in the direction corresponding to the higher resistance.

Note. This definition may not apply to tunnel diodes.

Floating voltage. Voltage between an open-circuited terminal and the reference point when a specified voltage is applied to any of the other terminals.
Breakdown voltage. Reverse voltage at which the reverse current through a junction becomes greater than a specified value.
Cut-off frequency. The frequency at which the modulus of a measured parameter has decreased to $1/\sqrt{2}$ of its low frequency value.

Note. For a transistor, the cut-off frequency usually applies to the short-circuit small-signal forward current transfer ratio for either the common-base or common-emitter configuration.

Temperatures

Case temperature. The temperature measured at a specified point on the case of a semiconductor device.
Thermal derating factor. The factor by which the power dissipation rating must be reduced with increase of ambient or case temperature.

SIGNAL DIODE AND RECTIFIER DIODE TERMS

Continuous (direct) reverse voltage. The value of the constant voltage applied to a diode in the reverse direction.
Peak reverse voltage. The highest instantaneous value of the reverse voltage occurring across a diode including all repetitive and non-repetitive transients.
Reverse current. The total conductive current flowing through the diode when specified reverse voltage is applied.
Forward voltage. The voltage across the terminals which results from the flow of current in the forward direction.
Forward current. The current flowing through the diode in the direction of lower resistance.
Differential resistance. The differential resistance measured between the terminals of the diode under specified conditions of measurement.
Forward d.c. resistance. The quotient of d.c. forward voltage across the diode and the corresponding d.c. forward current.
Reverse d.c. resistance. The quotient of the d.c. reverse voltage across the diode and the corresponding d.c. reverse current.
Small-signal capacitance. Differential capacitance at the diode terminals, measured under given bias conditions.
Reverse recovery time. The time required for current or voltage to a specified value after instantaneous switching from a specified forward current condition to a specified reverse bias condition.

TUNNEL DIODE TERMS

Forward direction. The direction of current flow within the diode for which the characteristic includes negative differential conductance.

Reverse direction. The direction of current flow within the diode for which the characteristic includes only positive differential conductance.

Peak point. The point on the characteristic corresponding to the lowest voltage in the forward direction for which the differential conductance is zero (see Fig. 9).

Peak point voltage. The voltage at the peak point (see Fig. 9).

Peak point current. The current value at the peak point (see Fig. 9).

Valley point. The point on the characteristic corresponding to the lowest voltage greater than the peak point voltage for which the differential conductance is zero (see Fig. 9).

Valley point voltage. The voltage value at the valley point (see Fig. 9).

Valley point current. The current value at the valley point (see Fig. 9).

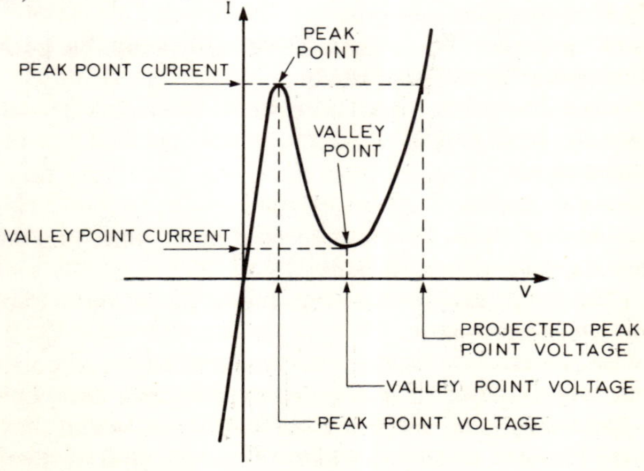

Fig. 9.—*Static voltage-current characteristic of a tunnel diode.*

Peak to valley point current ratio. The ratio of the peak point current to the valley point current.
Projected peak point. The point on the characteristic where the current is equal to the peak point current, but where the voltage is greater than the valley point voltage.
Projected peak point voltage. The voltage value at the projected peak point.
Negative differential conductance region. That part on the characteristic of a tunnel diode between the peak and valley points.
Case capacitance. The residual capacitance between the device terminals when the pn junction is not internally connected.
Series inductance. The total effective internal series inductance under specified conditions.

TRANSISTOR TERMS

General terms

Base terminal. The specified externally available point of connection to the base region.
Collector terminal. The specified externally available point of connection to the collector region.
Emitter terminal. The specified externally available point of connection to the emitter region.
Collector region. A region between the collector junction and the collector electrode of a transistor.
Emitter region. A region between the emitter junction and the emitter electrode of a transistor.
Base region. A region between the emitter and collector junctions of a transistor.
Collector junction. A junction between the base and collector regions normally biased in the reverse direction and through which the charge carriers flow from a region in which they are minority carriers to one in which they are majority carriers.
Emitter junction. A junction between the base and emitter

regions normally biased in the forward direction, and through which the charge carriers flow from a region in which they are majority carriers to one in which they are minority carriers.

Types of transistor

Junction transistor. Transistor having a base region and two or more junctions.

Note. The operation of a junction transistor depends upon the injection of minority carriers into the base region.

Bi-directional transistor. A transistor which has substantially the same electrical characteristics when the terminals normally designated as emitter and collector are interchanged.

Note. Bi-directional transistors are sometimes called symmetrical transistors. This term, however, is deprecated as it might give the incorrect impression of an ideally symmetrical transistor.

Tetrode transistor. A four-electrode transistor, usually a conventional junction transistor having two separate base electrodes and two base terminals.

Unipolar transistor. A transistor which utilises charge carriers of only one polarity.

Terms related to ratings and characteristics

Punch-through voltage. The value of the collector-base voltage above which the open-circuit emitter-base voltage increases almost linearly with increasing collector-base voltage.

Note. "Reach-through voltage" is a term also used in the U.S.A.

Saturation voltage. The residual voltage between collector and emitter terminals under specified conditions of base current and collector current, the collector current being limited by the external circuit.

Cut-off current (*reverse current*). Reverse current of the base-collector junction (or base-emitter junction) when the emitter or the collector is open-circuited, the reverse voltage being specified.

Emitter series resistance. The resistance between the emitter terminal and the internal inaccessible emitter point in an equivalent circuit.

Saturation resistance. The resistance between collector and emitter terminals under specified conditions of base current and collector current, when the collector current is limited by the external circuit.

Note. The saturation resistance may be determined either as the ratio of total voltage to total current or as the ratio of differential voltage to differential current; the method of determination must be specified.

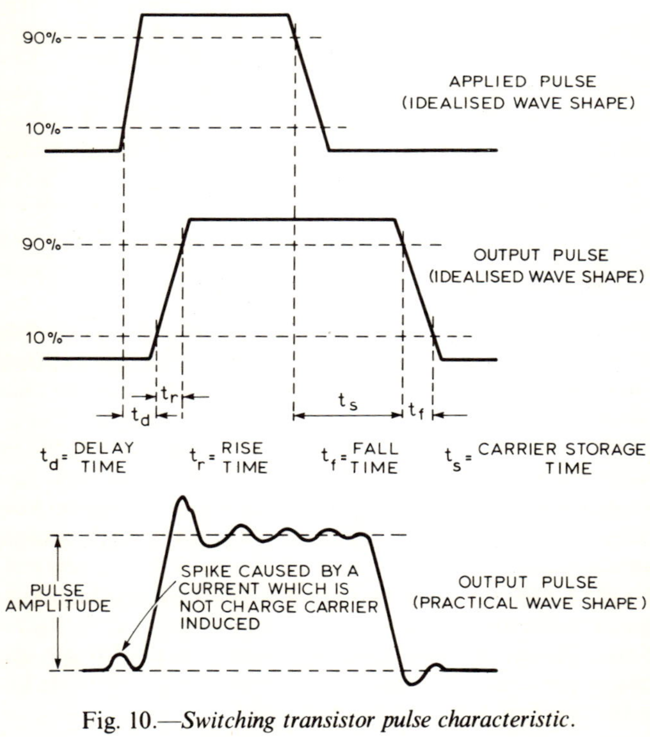

Fig. 10.—*Switching transistor pulse characteristic.*

Extrinsic base resistance. The resistance between the base terminal and the internal inaccessible base point in an equivalent circuit.

Collector depletion layer capacitance. The part of the capacitance across a collector-base junction that is associated with its depletion layer.

Note. The collector depletion layer capacitance is a function of the total potential difference across the depletion layer.

Delay time (*of a switching transistor*). The time interval between the application at the input terminals of a pulse which is switching the transistor from a non-conducting to a conducting state, and the appearance at the output terminals of the pulse induced by the charge carriers.

Note. The time is usually measured between points corresponding to 10 per cent of the amplitude of the applied pulse and of the output pulse, respectively (see Fig. 10).

Rise time (*of a switching transistor*). The time interval between the instants at which the magnitude of the pulse at the output terminals reaches specified lower and upper limits respectively when the transistor is being switched from its non-conducting to its conducting state.

Note. The lower and upper limits are usually 10 per cent and 90 per cent, respectively, of the amplitude of the output pulse (see Fig. 10).

Carrier storage time (*of a switching transistor*). The time interval between the beginning of the fall of the pulse applied to the input terminals and the beginning of the fall of the pulse generated by charge carriers at the output terminals.

Note. The time is generally measured between the 90 per cent values of the two pulse amplitudes (see Fig. 10).

Fall time (*of a switching transistor*). The time interval between the instants at which the magnitude of the pulse at the output terminals reaches specified upper and lower limits respectively when the transistor is being switched from its conducting to its non-conducting state.

Note. The upper and lower limits are usually 90 per cent and 10 per cent, respectively, of the amplitude of the output pulse (see Fig. 10).

Small-signal short-circuit forward current transfer ratio. The ratio between the alternating output current and the small sinusoidal input current producing it under small-signal conditions, the output being short-circuited to a.c.

Static value of the forward current transfer ratio. The ratio between the continuous (direct) output and the continuous (direct) input current, the output voltage being held constant.

Small-signal—open-circuit reverse voltage transfer ratio. The ratio of the alternating voltage appearing at the input terminals, when they are a.c. open-circuited, to the alternating voltage applied to the output terminals, under small-signal conditions.

Circuit configuration

Common base. Circuit configuration in which the base terminal is common to the input circuit and to the output circuit and in which the input terminal is the *emitter* terminal and the output terminal is the *collector* terminal.

Inverse common base. Circuit configuration in which the base terminal is common to the input circuit and to the output circuit and in which the input terminal is the *collector* terminal and the output terminal is the *emitter* terminal.

Common collector. Circuit configuration in which the collector terminal is common to the input circuit and to the output circuit and in which the input terminal is the *base* terminal and the output terminal is the *emitter* terminal.

Inverse common collector. Circuit configuration in which the collector terminal is common to the input circuit and to the output circuit and in which the input terminal is the *emitter* terminal and the output terminal is the *base* terminal.

Common emitter. Circuit configuration in which the emitter terminal is common to the input circuit and to the output circuit and in which the input terminal is the *base* terminal and the output terminal is the *collector* terminal.

Inverse common emitter. Circuit configuration in which the emitter terminal is common to the input circuit and to the output circuit and in which the input terminal is the *collector* terminal and the output terminal is the *base* terminal.

DEFINITIONS OF SYMBOLS

C_{ibo}	Open-circuit input capacitance, common base
C_{ibs}	Short-circuit input capacitance, common base
C_{ieo}	Open-circuit input capacitance, common emitter
C_{ies}	Short-circuit input capacitance, common emitter
C_{obo}	Open-circuit output capacitance, common base
C_{obs}	Short-circuit output capacitance, common base
C_{oeo}	Open-circuit output capacitance, common emitter
C_{oes}	Short-circuit output capacitance, common emitter
f_{hfb}	Small-signal short-circuit forward current transfer ratio cutoff frequency (common base)
f_{hfe}	Small-signal short-circuit forward current transfer ratio cutoff frequency (common emitter)
f_T	Frequency at which small-signal forward current transfer ratio (common emitter) extrapolates to unity
h_{FB}	Static forward current transfer ratio (common base)
h_{fb}	Small-signal short-circuit forward current transfer ratio (common base)
h_{FC}	Static forward current transfer ratio (common collector)
h_{fc}	Small-signal short-circuit forward current transfer ratio (common collector)
h_{FE}	Static forward current transfer ratio (common emitter)
h_{fe}	Small-signal short-circuit forward current transfer ratio (common emitter)
h_{IB}	Static input resistance (common base)
h_{ib}	Small-signal short-circuit input impedance (common base)
h_{IC}	Static input resistance (common collector)
h_{ic}	Small-signal short-circuit input impedance (common collector)
h_{IE}	Static input resistance (common emitter)
h_{ie}	Small-signal short-circuit input impedance (common emitter)

h_{ob}	Small-signal open-circuit output admittance (common base)
h_{oc}	Small-signal open-circuit output admittance (common collector)
h_{oe}	Small-signal open-circuit output admittance (common emitter)
h_{rb}	Small-signal open-circuit reverse voltage transfer ratio (common base)
h_{rc}	Small-signal open-circuit reverse voltage transfer ratio (common collector)
h_{re}	Small-signal open-circuit reverse voltage transfer ratio (common emitter)
I_B	Base current, d.c.
I_C	Collector current, d.c.
I_{CBO}	Collector cutoff current, d.c., emitter open
I_{CEO}	Collector cutoff current, d.c., base open
I_{CER}	Collector cutoff current, d.c., with specified resistance between base and emitter
I_{CES}	Collector cutoff current, d.c., with base short circuited to emitter
I_{CEV}	Collector cutoff current, d.c., with specified voltage between base and emitter
I_{DSS}	Drain current, d.c., with gate shorted to emitter
I_E	Emitter current, d.c.
I_{EBO}	Emitter cutoff current, d.c., collector open
I_{GSS}	Gate leakage current
r_{CE}(sat)	Collector-to-emitter saturation resistance
$V_{(BR)CBO}$	Breakdown voltage, collector-to-base, emitter open
$V_{(BR)CEO}$	Breakdown voltage, collector-to-emitter, base open
$V_{(BR)CER}$	Breakdown voltage, collector-to-emitter, with specified resistance between base and emitter
$V_{(BR)CES}$	Breakdown voltage, collector-to-emitter, with base short-circuited to emitter
$V_{(BR)CEX}$	Breakdown voltage, collector-to-emitter, with specified circuit between base and emitter
$V_{(BR)DGO}$	Breakdown voltage, drain-to-gate, source open

$V_{(BR)DSS}$	Breakdown voltage, drain-to-source, gate shorted to source
$V_{(BR)EBO}$	Breakdown voltage, emitter-to-base, collector open
$V_{(BR)GSS}$	Breakdown voltage, gate-to-source, drain shorted to source
V_{BC}	Base-to-collector voltage, d.c.
V_{BE}	Base-to-emitter voltage, d.c.
V_{CB}	Collector-to-base voltage, d.c.
V_{CE}	Collector-to-emitter voltage, d.c.
V_{CEO}	Collector-to-emitter voltage, d.c., with base open
$V_{CEO(SUS)}$	Collector-to-emitter (breakdown) sustaining voltage
V_{CER}	Collector-to-emitter voltage, d.c., with specified resistor between base and emitter
$V_{CER(SUS)}$	Collector-to-emitter (breakdown) sustaining voltage
V_{CES}	Collector-to-emitter voltage, d.c., with base short-circuited to emitter
$V_{CES(SUS)}$	Collector-to-emitter (breakdown) sustaining voltage
V_{CEX}	Collector-to-emitter voltage, d.c., with specified circuit between base and emitter
$V_{CEX(SUS)}$	Collector-to-emitter (breakdown) sustaining voltage
V_{CE}(sat)	Collector-to-emitter saturation voltage, d.c.
V_{EB}	Emitter-to-base voltage, d.c.
V_{EC}	Emitter-to-collector voltage, d.c.
V_{RT}	Reach-through voltage

INDEX